Role of Classical Mutation Breeding in CROP IMPROVEMENT

The Author

Dr. S.K.Datta obtained his B.Sc. Spl. Hons.(Botany) degree in 1968 and M.Sc. (Botany) with specialization in cytogenetics and plant breeding) degree in 1970 from Kalyani University, West Bengal. He started his research carrier in the field of induced mutagenesis for crop improvement at Bose Institute, Kolkata since April 1971 and obtained his Ph.D. degree in 1976 from Calcutta University for his thesis entitled "Cytogenetical studies in *Trichosanthes anguina* L.". Dr. Datta joined National Botanical Research Institute, Lucknow in 1977 and started his further research activities on improvement of different ornamental crops through induced mutagenesis. He was Head, Mutation Breeding Laboratory, NBRI since 1985. He has developed more than 80 new varieties in different ornamentals for floriculture trade through induced mutation. He was awarded D.Sc. degree in Botany for his extensive work on ornamentals. Dr. Datta has been associated with several professional societies and committee.

Dr. Datta has published over 175 research papers on mutation breeding in National and International journals. He has published one book "Ornamental Plants: Role of mutation" and five Bulletins. He has contributed chapters on mutation breeding in several books and published series of review research papers on induced mutagenesis. He was invited to present his paper on mutation breeding at Tsukuba, Japan during the 6th International Congress of the Society for the Advancement of Breeding Research in Asia and Oceania (1989); at Vienna, Austria during International Symposium on the contribution of Plant Mutation Breeding to crop improvement, 1990; at Vienna, Austria, during FAO/ IAEA International Symposium on the use of induced mutation and molecular techniques for crop improvement, 1995; at Kuala Lumpur, Malaysia, during International Nuclear Conference 2002.

Dr. Datta was also deputed to Berlin under CSIR-DAAD Scientists Exchange Programme for two months (1992). Considering the quantum of work done and published literature on induced mutagenesis, International Atomic Energy Agency (IAEA), Vienna selected. Dr. Datta as "Expert on Mission" for evaluation of mutation breeding project sponsored by IAEA to Philippines (October 2001); and Jakarta, Indonesia for project evaluation mission.

Dr. Datta is now Deputy Director, Coordinator and Head, Botanic Garden and Floriculture, NBRI and he is working on different aspects related to cytomorphology, *in vivo* and *in vitro* induced mutagenesis, conventional breeding, tissue culture, management of chimera, conservation, dehydration of flowers and floral crafts etc. on different ornamentals like Amaryllis, Asiatic Hybrid Lily, Bougainvillea, Chrysanthemum, Dahlia, Gerbera, Marigold, Gladiolus, Hibiscus, Rose, Tuberose, *Lantana depressa*, Orchid etc.

Role of Classical Mutation Breeding in CROP IMPROVEMENT

— *Editor* —

S.K. Datta

Scientist 'F' & Area Coordinator, Botanic Garden & Floriculture
National Botanical Research Institute
Lucknow – 226 001 (U.P.)

2005

DAYA PUBLISHING HOUSE

Delhi - 110 035

ISBN 81-7035-354-8

Published by : **Daya Publishing House**
1123/74, Deva Ram Park
Tri Nagar, Delhi - 110 035
Phone: 27383999
Fax: (011) 23244987
e-mail : dayabooks@vsnl.com
website : www.dayabooks.com

Showroom : 4762-63/23, Ansari Road, Darya Ganj,
New Delhi - 110 002
Phone: 23245578, 23244987

Laser Typesetting : **Classic Computer Services**
Delhi - 110 035

Printed at : **Chawla Offset Printers**
Delhi - 110 052

PRINTED IN INDIA

Foreword I

Mutation breeding is now a well established method for crop improvement. Mutation has already been recognized as potential technique for crop improvement since the discovery of mutation effects of radiations on plants (Gager 1908, Muller 1927, Stadler 1928). Monumental literature have been gathered on the basis of world wide activities on mutagenesis. More than 2200 new crop varieties have been developed through induced mutagenesis and commercialized world wide. Compilation of literature on impact of classical mutation breeding for crop improvement is an immensely valuable effort.

I am happy that Dr. S.K. Datta, Scientist F, Coordinator and Head, Botanic Garden and Floriculture, National Botanical research Institute, Lucknow, India has taken initiative to compile the research results of classical mutation breeding achieved so far on different crops. Dr. Datta is an internationally well acclaimed expert in mutation breeding. He has been successfully employing this technique during the past 30 years in genetic enhancement of a variety crop plants like vegetables, pulses, medicinal aromatic, fuel oil and ornamental plants. The book that Dr. Datta compiled presents the results of important achievements on induced mutagenesis in different countries.

I hope that the book will serve as a valuable reference book to all those who are interested in induced mutagenesis research.

Dr. P. Pushpangadan
Director
National Botanical Research Institute
Lucknow, India

Foreword II

Domestication of crop plants was intimately associated with the selection of the mutants by the pioneering farmers. Present day free threshing wheat, non-shattering rice; hull-less barley, and maize cobs are some of the well-known examples of single or multiple mutations that have grossly changed these crops from their wild ancestors. Naturally occurring mutations selected since domestication contribute to the genetic diversity that the plant breeders exploit to develop crop cultivars of modern agriculture. Biodiversity of the vegetation-conservation of which is currently emphasized in all flora, represents mutations that could compete with the wild types, and survive in natural habitats. Errors in DNA replication that escape "proof reading" are mainly responsible for the spontaneous mutations. Muller in 1927 discovered that mutations can be induced in fruit fly (*Drosophila melanogaster*) by exposing them to X-rays. Stadler obtained similar results in 1928 after exposing barley and maize seeds. Since then, a wide range of radiations, and chemicals, some of them commonly used in industry, and agriculture are now known to induce mutations in different test systems.

Immediately after these discoveries, plant breeders initiated wide-ranging experiments to induce, and select desired mutations of agronomic value in different crop and ornamental plants. The method came to be known as Mutation Breeding. It is an established method for the improvement of plants now. On the basis of worldwide activities on induced mutagenesis enormous quantity of literature has been generated. More than 2250 induced mutation-derived varieties have been commercialized globally. Compilation of literature on the impact of induced mutations for crop improvement was necessary to provide a perspective at this time when the interest is rapidly shifting to more expensive molecular tools such as "genetic engineering" for the improvement of cultivated plants. The recombinant DNA methods facilitate transfer of the gene mutations from microbes, plants and animals that were previously outside of the crop species gene pool. The future of induced-mutagenesis is in isolation, cloning, *in-vitro* modification of the desired genes, and transferring them back into plant genomes. Till such molecular methods are in their easy reach, classical methods of induced mutagenesis involving exposure to radiations, or treatment with chemical mutagens, will remain a highly cost-effective tool in the hands of plant breeders.

I am happy that Dr. S.K. Datta, Deputy Director, Co-ordinator and Head, Botanic Garden and Floriculture, National Botanical Research Institute, Lucknow, India has taken the initiative to compile the research results of classical mutation breeding achieved so far on different crops. The book presents a state-of-the-art of the advances of mutation breeding in crop improvement.

I hope that the book will be a useful reference material for all those interested in induced mutagenesis, and its applications for the genetic enhancement of cultivated plants.

Dr. C.R. Bhatia
Secretary, Govt. of India (Retired)
Department of Biotechnology
Ministry of Science and Technology
New Delhi

Preface

The concept of induced mutagensis for crop improvement developed dated back to the beginning of the 20th century. Informations accumulated on optimal treatment doses, treatment condition, mutation frequency and mutation spectra using ionizing and non-ionizing radiations, and chemical mutagens. High potential for bringing genetic improvement by induced mutation was realized.

Induced somatic mutation breeding holds promise for effective improvement and have high potential for bringing about genetic improvement. Mutation techniques by using ionizing radiations and other mutagens have successfully produced quite a large number of new promising varieties in different crops world wide. Monumental literature have been gathered on the basis of world wide activities on mutagenesis.

Extensive research activities on classical induced mutagensis are going on at Floriculture Laboratory, National Botanical Research Institute, Lucknow, India for the last 30 years. The main research activities are to develop new and novel ornamental varieties using gamma radiation. Research carried out covers radiosensitivity, selection of materials, method of exposure, type of irradiation/ mutagens, detection of mutations, isolation of mutants, multiplication of mutants, management of chimera, evaluation of mutant for commercial exploitation etc.

The results and mutants obtained on different crops world wide using induced mutagensis are of very interesting both from academic and applied point of view. There is technological advancement in crop improvement. Recently application of biotechnology to the development on new varieties in different crops have been seriously considered. Recent advancement in genetic engineering technology have opened up the possibility of crop improvement in a highly controlled manner. The possibility hold the promise of generating a much wider desirable variability than in now available to the breeder. Biotechnology offers new and exciting challenges for the future. But it just enables researchers to start unraveling the molecular details and understanding of the function and interaction of individual genes for successful application of genetic engineering in crop improvement.

Present status of all the techniques clearly indicate that induce mutagenesis technique is well standardized whereas the molecular technique is on progress. At this stage induced mutation breeding combined with *in vitro* chimera management has tremendous potential for crop improvement.

Attempts have been made in the present book to compile the work carried out on different crops for their improvement in different countries. It will be a useful exercise to compile available research results on induced mutagenesis on different crops under the title " Role of classical mutation breeding in crop improvement". Chapters contributed by multidisciplinary group of competent scientists will be an excellent reference book on classical mutation breeding.

It is hoped that the book will prove immense value to research workers, teachers, students and individuals who are interested for crop improvement through induced mutagenesis.

The book is most respectfully dedicated to my parents, whose unfathomable love kept me ever up and doing.

I sincerely acknowledge my long association at National Botanical Research Institute, Lucknow, India where did maximum induced mutagenesis work on different crops.

I wish to record my sincere thanks to Dr. P. Pushpangadan, Director, NBRI, who always encouraged and appreciated my activities on induced mutagenesis. I heartily acknowledge the warmth of my wife who always stood by me stead fast. I am gratefully indebted to all my laboratory and experimental field colleagues for their profound affection, encouragement and ready help during the various phases of the work.

S.K. Datta
Botanic Garden and Floriculture
National Botanical Research Institute
Rana Pratap Marg
Lucknow – 226 001, U.P., India
E-mail: subodhdatta@usa.net
subodhskdatta@rediffmail.com
subodhskdatta@yahoo.com

Contents

Chapter 1

Use of Induced Mutations for Crop Improvement: Revisited

H.Yamaguchi

Department of Radiology, Junior College, Komazawa University Komazawa, Tokyo – 154, Japan

ABSTRACT

The usefulness of mutation breeding in cr"P improvement depends on the efficiency for producing genetic variation and on the availability of effective techniques for detecting and isolating mutants. Fundamental aspects of experimental mutagenesis and mutant-isolating techniques are reviewed. The induction of increased genetic variation is feasible with interference into the processes of DNA repair, use of hybrid seeds for mutation induction and application of mutagenesis in plant tissue culture. The usefulness of ion beam as a mutagen was pointed out because its different genetic effects from gamma-rays. Finally, some new research directions were discussed on the integration of plant molecular biology into the mutagenesis and detection of induced mutants.

INTRODUCTION

In attempts to induce useful phenotypic variation in plants, mutagenic treatments have been used for more than 70 years. A lot of mutant lines have been isolated by the mutagenic techniques developed during this periods and extensively used in many different fields of plant genetic research and crop breeding. Fundamental and applied aspects of experimental mutagenesis have been extensively reviewed (Gottschalk1983, Yamaguchi 1991, Brunner1991).

The number of genes expressed during the lifetime of a plant is estimated to be between 16,000 and 33,000 (Gibson and Somerville, 1993). Spontaneous and induced mutations in plants contribute for genetic dissection of the function of wild-type genes. In Arabidopsis and other well established genetic model species, namely barley, maize, rice or tomato the chromosomal location of new gene mutations has been mapped due to a cause-and-effect relationship between changes in genotype and alterations in phenotype, and it has opened the way for the construction of high-density, integrated genetic maps based on different types of DNA markers covering much of the genome.

As a result of DNA construct (transgene) incorporation, host DNA near the site of integration frequently undergoes various forms of sequence duplication, deletion, or rearrangement. Such alterations, if sufficiently drastic, may disrupt the function of normally active host genes at the insertion site and constitute insertional mutagenesis resulting an aberrant phenotype. Such events cannot be purposefully designed, but they had led to the serendipitous finding of unsuspected genes and gene functions.

Molecules inhibiting transcription or translation can be designed on a rational basis. Because of the general base-pairing rules, intervention using oligonucleotide antagonists is a universal approach. Since a report of Zamecnik and Stephenson (1978) who proposed the use of oligonucleotides directed against complementary viral nucleic acid sequences for inhibition of virus replication, therefore, a great deal of work has been devoted to this principle. Antisense oligonucleotides are synthetic oligo-nucleotides that bind to certain complementary regions of the mRNA, thereby inhibiting protein biosynthesis. For example, alteration of polygalacturonase (PG) gene function as a result of transforming tomato plants with antisense PG genes resulted in fruit that soften not and less quickly (Gray *et al.*, 1992). This change is similar as the induced mutation lacking the normal activity of the gene, and such a change will be possible to produce also with induction of mutations.

Insertional mutagenesis or gene-tagging, based on the activity of endo-genous transpositions is becoming the important techniques in plant development biology. Despite the success of insertional mutagenesis in Arabidopsis, however, the question still remains as to whether such approaches are possible into every species. For example, insertion mutations can not be obtained due to the lethal nature of the null phenotype of essential genes (Mouvad *et al.*, 1994). In contrast, classical point mutants can be easily detected due to resistant or leaky alleles of essential genes (Drake 1970). Taking together, therefore, the approaches of classical and insertional mutagenesis complement each other for molecular dissection of plant gene and its application to crop breeding in the future.

Methods and Types of Mutagens

In addition to that seeds, pollen and various tissues and cells have been employed for the production of mutations with various mutagens, tissue–culture-based mutagenesis has been used to produce somaclonal variation, which can induce a range of chromosomal changes, as well as gene mutations (Evans and Sharp 1986).

Classical mutagenesis has been generated with the use of physical mutagens, namely ionizing radiations (X-rays, gamma rays, neutrons and ion beams) and UV light, and a variety of chemical agents. Genetic effects with experimental mutagenesis includes point mutations, such as base substitutions at different sites in a particular gene, from complete loss of gene function to alteration of levels or timing of mRNA. expression, or changes in encoded protein activity. Of point mutations, deletions of some bases in a gene are more robust in effect and will often result in null mutations. Ionizing radiation causes single–and double-strand breaks, which result in chromosomal rearrangements or gene mutations, following DNA repair. The most commonly used chemical mutagen for seed treatment is ethyl methane sulfonate (EMS), which alkylates DNA, causing base pair transition, mostly GC to AT, due to the mispairing of O^6-alkyl-G with T. It is assumed, therefore, that EMS mutagenesis results in the transition-type of point mutations.

Highly efficient methods of radiation and chemical mutagenesis with use of seeds have been developed. Mutagens and methods for their use in plant has been published by the International Atomic Energy Agency (1977). A number of factors alter the mutagenic response of plants in effectiveness, mutations per unit dose, and efficiency, ratio of mutations to damage, and they are dependent on the specific properties and treatment conditions of mutagens, and on the size, the nucleotide composition and genomic location of target gene, chromatin structure during a cell cycle and efficiency of DNA repair. The mechanism that recognizes and eliminates an aberrant cell has been termed 'cellular proofreading', because the mistake is erased rather than repaired. The term also draws attention to the fact that apoptosis is just an effective step (Brash, 1996).

Correlation of the S Phase with Mutation Fixation

Mutation frequencies are reduced (often to background levels) if mutagen-treated cells are prevented from entering S phase for a sufficient period. This has been achieved by maintaining confluence (Grosovsky and Little1983, Maher *et al.*, 1979) or by synchronization with hydroxyurea (Stone-Wolff and Rossman 1982) or low temperature (Ernings *et al.*, 1985). This was also confirmed in the mutagenesis of plant materials.

From the results of mutations induced by tritiated thymidine treatment of soaking rice seeds, it was assumed that *de novo* synthesis of DNA in the cell from which germinal tissue originate starts from about 60 hr after soaking (Yamaguchi1969a). In gamma irradiation, definite peak in lethality or sterility and chlorophyll mutation was obtained 58 hr after seed soaking. Irradiation of 60 hr soaked seed, however, produced little if any mutation (Yamaguchi, 1969b). After ethyl methanesulfonate (EMS) treatment of rice seeds, meanwhile, the mutation frequencies were high at 56 hr soaking (late Gl), decreased suddenly at 58 hr soaking (middle S), increased again to a note worthy value at 60 hr soaking (late S and/or early G2) and rapidly decreased later than 62 hr after seed soaking (Yamaguchi and Matsubayashi1973). These results suggest that mutations were induced very efficiently after mutagenic treatment of the apical meristem in late Gl to early S.

DNA Damage and Mutational Mosaics

Ionizing radiation and chemical mutagen damage the DNA in every cell. These damages include single–and double-strand breaks, DNA-protein cross-links, Sl endonuclease-sensitive sites indicative of clustered base damage, and single-base damage such as 8-hydroxypurine, formamidopyrimidine, and thymine or cytosine glycol. They interrupt the continuity of genetic information by inhibiting or preventing transcription, DNA replication, and cell division.

When the nuclei isolated from dormant barley seeds immediately after–irradiation were lysed on the top of sucrose gradients, sedimentation velocity studies showed that single-strand breaks occur in the nuclear DNA and that single-strand breaks per g DNA increase linearly with radiation dose (EI-Metainy *et al.*, 1971). Along with single-strand breaks of DNA, inorganic phosphate was liberated and phosphomonoester groups were formed. When the nature of thc 5'-cnd groups in single strand breaks of DNA was determined by the polynucleotide kinase technique, about 91 per cent of the total S'-termini was found to carry 5'-phosphoryl termini. These termini may originate from an oxidation of the C'-3 of the deoxyribose and a subsequent splitting of the phosphoester bond. In comparing the amount of phosphomonoester groups formed with that of the 5'-phosphoryl termini, it was estimated that only about 40 per cent of the total phosphomonoester groups formed are S'-phosphomonoesters (EI-Metainy *et al.*, 1973). One cannot ruled out whether all the other phosphomonoesters are 3'-phosphomonoesters, as being shown that radical attack on the C atom of the sugar moiety results in the formation of a labile phosphate group with the eventual release of inorganic phosphate.

Muller *et al.* (1961) suggested that mutational mosaics would be expected frequently if X-rays affected only one DNA strand and replication was carried out semi-conservatively. Mosaics were more frequent after X–or gamma irradiation of rice seeds with lower doses rather than with higher doses and a high frequency of mosaics was detected after the treatment of rice seeds with ethyleneimine (Yamaguchi and Jalil Miah 1964).

Mutations Induced as DNA Repair Errors

Typically, single-strand breaks cannot be rejoined by one-step ligation reactions since ionizing radiation frequently results in damaged 3' termini and often in a single-base gap rather than a simple nick. Hence, nonrecombinational repair requires the action of multiple enzymes involved in base excision repair (Yamaguchi *et al.*, 1976). To obtain information on the refractoriness of mutation induction, the 58 hr soaked seeds of rice were treated for 2 hr with EMS solution and centrifuged for 15 min before termination of the EMS treatment. About three times increase of the mutation frequency was found with post–treatment. Also, the evidence for the existence of post-treatment processes in mutation induction came from an experiment with post–treatment of 5-bromodeoxyuridine (BUdR). When BUdR was administered for one hour immediately after diethylsulfate treatment of barley soaked seeds, the increase in mutations was found after post-treatment with BUdR. Since BUdR is incorporated in place of thymidine at DNA synthesis, an increase of mutations with the post-treatment is probably through errors of DNA repair (Yamaguchi 1973). Thus, it was suggested that the refractoriness to mutation induction observed in middle S was the result of post-treatment repair process(es) that remove most of the initial damage induced by mutagenic treatment.

Mutation frequencies increased remarkably with post-treatments of caffeine, EDTA and BUdR if each presented during the first 5 hours of germination of-irradiated barley seeds (Yamaguchi, 1979). This suggests that interference in the process(es) of unscheduled DNA repair synthesis (Yamaguchi *et al.*, 1975) results in more mutations.

Postreplication repair is operationally defined as the sum of processes that restore high-molecular weight DNA during or after replication of a mutagen-treated template. A model of postreplicational repair takes into account the ability of RecA protein to catalyze pairing and strand exchange reactions between homologous DNA molecules (Howard-Flanders *et al.*, 1968, 1984, West *et al.*, 1981).

Mitotic recombinational mechanisms appear to be the only way of overcoming DNA double-strand breaks induced by ionizing radiation. The recombinational nature of the underlying repair process requires the presence of two DNA copies, as indicated by the relative X-ray resistance of haploid G2 and diploid wild-type cells (Mortimer 1958), or the synapsis of mitotic chromosomes, as suggested by the relative gamma ray susceptibility of partially asynaptic mutants (Yamaguchi, 1974). In *Saccharomyces cerevisiae,* RAD52-dependent homologous recombination is the major repair pathway for double-strand breaks. There is no indication of fundamental mechanistic differences in the various pathways of recombinational repair in higher eukaryotic cells, yeast cells, or even *Eschericia coli* (Friedberg *et al.*, 199S).

Since a recombination repair mechanism is suggested to be induced in cells containing damaged DNA, ionizing radiation may increase the genetic variation by somatic recombination in addition to induced mutation in the irradiation of FI seeds. Regarding plant yield of rice, the increase of genetic variance due to r-irradiation was significant in the hybrid, the predicted gain from selection was higher in the hybrid than the two parental varieties and indeed more desirable variants by the irradiation of hybrid seeds were possible to select in comparison to the irradiated seeds of pure variety (Jalil Miah and Yamaguchi I965).

High Let Effects in Mutation Induction

From comparisons of radiation types for absolute effectiveness per unit dose and dependence on modifying factors implying differences in the initial spectrum of molecular damage and in repair of the damage, it is shown that radiation track structure is a primary determinant of biological effectiveness.

For high LET radiation, it can be noted that it is not simple double-strand breaks but the non-reparable breaks that correlate well with high biological effectiveness (Eguchi-Kasai and Cox 1995). Consequently, it is suggested that the DNA repair system does not play a major role against the attack of high LET radiations.

DNA double-strand breaks produced by high LET radiation may not differ from those induced by low LET. The production of double-strand breaks was not remarkedly affected by the radiation type and LET, and in all cases an RBE not significantly higher than unit was observed (Prise *et al.*, 1992; Jenner *et al.*, 1992, Newman *et al.*, 1997). For mutation induction at x–linked hypoxanthine-guanine phosphoribosyl transferase *(hprt)* locus of cultured mammalian cells an increase of the RBE with LET was found up to 31keV/m (Belli *et al.*, 1991, 1993).

If double-strand breaks are the lesions responsible for mutation induction, the number of mutagenic events or double-strand break rise with LET. It appears that different types of radiation produce lesions of different quality, namely with different probabilities of biological consequences. There is a general agreement that increasing LET also increases the probability of producing clusters of lesions which, in turn, could result less reparable than the more scattered ones produced by low LET radiation (Goodhead, 1994).

Dry seeds were irradiated with carbon ions (LET:ll0 keV/?m) and two types of mutations were induced in Arabidopsis. One was a mutation of anthocyan accumulation in testa of seed and the other was a ultraviolet light B (UVB)-resistant mutation with high reparability of radiation damages. The former mutation was a change of the locus linked closely to *hga280* with 3.3 cM on the chromosome l (Tanaka *et al.*, 1996). When the explants of leaf and floral petals of Chrysanthemum plant were irradiated with carbon ions (LET:113 keV/?m), on the otherhand, novel mutations of flower color were found. Frequency of mutations induced by the ion beams was approximately half of that by gamma rays in both floral petal and leaf. Interestingly, however, most mutants had complex and stripe types, whereas gamma ray induced mutants had a single flower color (Nagatomi *et al.*, 1997). From these data, it is suggested that carbon ions are used as a mutagen with different action from -rays.

Genetic Nature of Induced Mutants

In plant, genomic and cDNA sequencing is under way for *Arabidopsis thaliana* and rice. Hofte *et al.* (1993) found that out of 1152 cDNAS of Arabidopsis 375 (32 per cent) were similar to known proteins. Of these 375 sequences, 35 per cent were identical to known Arabidopsis genes, and 33 per cent were similar to gene from other plants. The latter, in most cases, was considered to represent the homologous gene in Arabidopsis. However, 7 per cent of the clones sequenced corresponded to new members which had not been identified by classical methods, and 25 per cent to genes which had never been sequenced in plants. This indicates the existence of abundantly expressed genes or gene families which have not yet been identified in plants. These unidentified proteins included those which are only present in the particular tissue or culture conditions, those which play a more general role in cell function, and those which may be specific to plants. Expressed sequence tags (ESTs) of cDNAs for rice (Uchimiya *et al.*, 1992, Sasaki *et al.*, 1994) and

Arabidopsis (Hofte *et al.*, 1993, Newman *et al.*, 1994) are available in the EST databank, dbEST Rice was chosen as a model species because many morphological and isozyme markers have already been mapped to their respective chromosomes, in addition to smallest genome among the cereals and its agronomic importance. *Arabidopsis thaliana* was chosen as a model species for sequencing projects for its small genome, short generation time and the accumulated genetic loci marked by mutation. Sequence from the 5' extremity of ESTs allows to identify putative protein products, while 3' sequences can provide gene-specific probes. As rice and Arabidopsis represent the division of the plant kingdom into monocots and dicots, it is most probable that these sequences are at least equally well conserved in other plant species. For closely related organisms, therefore, it should be possible to use the ESTs as hybridization probes to detect homologous genes.

Detection and Isolation of Mutated Plants

The probability of isolating a particular mutation following mutagenic treatment is a product of the frequency of mutation and the frequency of mutant detection. One of the most important factors is the fate of a mutagen-treated shoot apical stem cell and the contribution of its cell lineage to the inflorescence of the mature plant (Stadler, 1930, MacKey 1954, Fujii 1960, Yamaguchi 1962). The number of shoot apical stem cells (genetically effective number) in rice seed has been estimated as 1-6 cells (Yamaguchi 1962, Osone 1963). Therefore, the frequency of appearance of mutants within individual M_1 panicle, namely individual M_2 families can range from 25 per cent in the former case to 4.2 per cent in the latter extreme.

The recovery of mutants induced by mutagens is limited by somatic effects, such as reduced viability and fertility. Any diplontic selection against the cellls carrying a mutation would decrease the frequency of mutant appearance (Gaul 1963, 1964). Also, variations in expression of mutant phenotypes and in the viability of mutants might influence the efficiency of mutant detection.

The structure of meristematic regions and the formation of new meristems from differentiated tissue is particularly important in the mutagenic treatment of vegetative organs. Treatment of vegetative organs results mostly in chimeras, as a consequence of the one-cell origin of a mutation and of the stable character of the apical layers. Hence, special isolation techniques have to be adopted, such as cutting-back methods, before solid mutants can be isolated (Bauer, 1957). One of the methods avoiding chimera formation is the adventitious bud technique (Broertjes *et al.*, 1968). This technique is based on the phenomenon that the apex of adventitious buds originates from one epidermal cell only.

Irradiation of buds with relatively high dose reduced the number of stem cells responsible for shoot development to one or very few out of which non-chimeric shoots can be recovered (Sekiguchi *et al.*, 1971). Significant progress has been achieved in many non-chimeric technologies based on cellular totipotency of plant cell and tissue culture. Cells or tissues can be treated with mutagens either before isolation and explanting or when materials are already in culture. For example, Nagatomi *et al.* (1996) irradiated the explants from leaf, floral bud and flower petal of Chrysanthemum plant after 3 days-incubating on the callus induction medium, transferred to a new callus induction medium immediately after irradiation, subcultured the induced calli on regeneration medium and obtained non-chimeric mutants of the regenerated plants. The combined method of irradiation and tissue culture was more effective for the production of flower color mutations, yielding their frequencies ten times higher when compared with the conventional chronic cutting method.

Multiple mutations might influence the phenotypic expression of an aimed mutation and complicate further analysis. If the new mutation and the contaminating mutation are unlinked or linked but lie more than 50cM apart, six backcrosses will give a 98 per cent probability of separating the new mutation from the contaminating mutation. The problem can be resolved according to recombination between the new gene and the contaminating mutation, and the allelism of the new mutation.

Genetic Analysis of a New Mutation

Genetic analysis begins with a study of the mutant phenotype and its inheritance through subsequent generations, tests for allelism with known mutants which show a similar phenotype, reciprocal crosses to check the possibility of non-nuclear location of the mutation, segregation analysis and cosegregation analysis with various tester lines. By this sequencial tests the nature of the genetic factor causing particular changes in phenotype becomes to be identify.

Visible mutants were used to establish the principle of genetic mapping, based on the hypothesis that likelihood of cotransmission of any two markers reflected the proximity between the markers along the chromosome (Mather 1957). The localization of a new mutation to a particular chromosome is possible with the use of trisomics. In this case, chromosome assignments can be made with near certainty and small population sizes. Alternatively, reciprocal translocations can be employed. In addition to these classical, scheme, molecular markers, such as restriction fragment length polymorphisms (RFLPs) (Borstein *et al.*, 1980), random amplification of polymorphic DNA (RAPDs) (Williams *et al.*, 1990), cleaved amplified poly-morphic sequences (CAPSs) (Konieczny and Ausubel 1993, Jarvis *et al.*, 1994), and simple sequence length polymorphisms (SSLPs) (Tautz 1989) help to map a new mutation into a particular chromosome. The advantage of molecular markers is that they do not interfere with the expression of a new mutation and further analysis of DNA from the same F2 populations is possible whenever new molecular markers become available. Large numbers of mapped molecular markers are available in many species.

Once linkage has been established, fine mapping of a mutation will be required as a prerequisite for map-based gene cloning. Also, small and large deletions have proved to be useful for fine-scale deletion mapping and analysis of function, along particular chromosomal regions. Genetic mapping provides the information needed to implement DNA marker-assisted selection. DNA markers diagnostic of traits which are difficult to measure is of particular potential value to the crop breeder.

Classical quantitative genetic theory has suggested that complex traits such as yield or seed size may be influenced by a virtually infinite number of genes, each with a very small effect. These genes are commonly known as polygenes (Mather and Jinks 1971). Over the past 20 years, punctuational models which invoke more rapid selection for few genes with larger effects have gained support. Many important aspects of dichotomy are still awaiting resolution.

The mutation in polygenes is known as micromutation, and its usefulness in crop breeding has been emphasized by several workers (Lawrence 1965, Gaul 1966, Scossiroli 1966), because complex traits are important in agricultural plants. Experiments demonstrated that random mutations in quantitative traits could be induced in both positive as well as negative directions with the increase in variances. According to heritability studies, such changes is due to increased genetic variation in the population. There are, however, conflicting reports whether induced mutations equally in plus and minus directions or it is undirectional. Jalil Miah and Yamaguchi (1964) observed in gamma irradiated progenies of rice that without selection the mean values for seed size decreased due to successive irradiation and highly significant in breadth. Successive irradiation with selection shifted the mean values towards the aimed direction.

If independent mutations in a virtually infinite number of genes with a very small effect could confer the phenotypes studied, the correspondence we observed would be unlikely occur. Recent advances in the molecular dissection study of complex phenotypes enable to do the experiments of quantitative trait loci (QTLS) mapping (Paterson 1998). The nonrandom distributions of QTLs tend to support punctuational models.

Future Prospects

A variety of techniques are now available for genome-wide screening of the mutants, such as alteration in copy number, structure, and expression of genes and DNA sequences. Comparative genomic hybridization (CGH) which is based on a modifies *in situ* hybridization, was the first molecular cytogenetic tool that allowed comprehensive analysis of the entire genome and has now become one of the most popular genome scanning techniques (Kallioniemi *et al.*, 1992, du Manoir *et al.*, 1993). In CGH, differentially labeled test and reference DNAs are co-hybridized to mutant metaphase spreads. CGH allows screening for DNA sequence copy-number changes, and provides a map of those chromosomal regions that are gained or lose in a DNA specimen.

Combining with the completion of the rice genomic sequence and a catalog of all genes in the not-so-distant future, the developments in this field are likely to be surprisingly fast.

REFERENCES

Bauer, R. 1957. The induction of vegetative mutations in *Ribes nigrum*. Hereditas: 33-337.

Belli, M., Cera, F., Cherubini, R., Ianzini, F., Moschini, G., Sapora, O., Simone, G., Tabocchini, M. A., and Tiveron, P. 1991. Mutation induction and RBE -LET relationship of low-energy protons in V79 cells. Int. J. Radiat. Biol. 59: 459-466.

Belli, M., Cera, F., Cherubini, R., Haque, A. M. I., Ianzini, F., Moschini, G., Sapora, O., Simone, G., Tabocchini, M. A. and Tiveron, P. 1993. Inactivation and mutation induction in V79 cells by low energy protons: Re-evalua–tion of the results at the LNL facility. Int. J. Radiat. Biol. 63: 331 -337.

Borstein, D., WhiLe, R. L., Skolnick, M. and Davis, R. W. 1980. Construction of a genetic linkage map in man using restriction fragment length polymorphisms. Am. J. Hum. Genet. 32: 314-331.

Brash, D. E. 1996. Cellular proofreading. Nature Med. 2: 577.

Broertjes, C., Haccius, B. and Weidlich, S. 1968. Adventitious bud formation on isolated leaves and its significance for mutation breeding. Euphytica 17: 321-344.

Brunner, H. 1991. Methods of induction of mutations. Mandel, A. K., Ganguli, K. K. and Banerjee, S. P. (eds). Advances in Plant Breeding, Vol. 1: 187–220.

Drake, J. W. 1970. The Molecular Basis of Mutation. Holden-Day, San Francisco.

duManoir, S., Speicher, M. R., Joos, S., Schroeck, E., Popp, S., DOehner, H., Kovacs, G., Robert-Nicoud, M., Lichter, P. and Cremer, T. 1993. Detection of complete and partial chromosome gains and losses by comparative genomic *in situ* hybridization. Hum. Genet. 90: 590-610.

Eguchi-Kasai, K. and Cox, A. B. 1995. Radiation quality and the relationship among cellular radiosensitivity, DNA repair and RBE. Proc. 10th Intern. Cong. Radiat. Res. Vol. 2: 118-121. EI-

Metainy, A., Takagi, M., Tano, S. and Yamaguchi, H. 1971. Radiation-induced single-strand breaks in the DNA of dormant barley seeds. Mutat. Res. 13: 337-344.

El-Metainy, A., Tano, S., Yano, K. and Yamaguchi, H. 1973. Chemical nature of radiation-induced single-strand breaks in the DNA dormant barley seeds *in vivo*. Radiat. Res. 55: 34-333.

Ernings, I., Groenendijik, R. T. L., van Zeeland, A. A. and Simons, J. W. I. M. 1985. Differential response of human fibroblasts to the cytotoxic and mutagenic effects of UV radiation in different phases of the cell cycle. Mutat. Res. 148: 119-128.

Evans, D. A. and Sharp, W. R. 1986. Applications of somaclonal variation. Bio/Technology 4: 528-532.

Friedberg, E. C., Walker, G. C. and Siede, W. 1995. DNA Repair and Mutagenesis. American Society for Microbiology, Washington, D.C.

Fujii, T. 1960. Mutations in einkorn wheat induced by X-rays. VI. Segregation ratio and viability of several chlorophyll mutants. Seiken Ziho 11: 12-20.

Gaul, H. 1963. Mutationen in der Pflanzenztichtung. Z. Pflanzenz. 50: 194-307.

Gaul, H. 1964. Mutations in plant breeding. Radiat. Botany 4: 155-232.

Gaul, H. 1966. Ztichterische Bedeutung von Kleinmutationen I. Durch Rontgen–strahlen induzierte Variabilitat van Kornertrag, Korngrosse und Vegetationslange bel der Gerste Haisa II. Z. Pflanzenz. 55: 1-20.

Gibson, S. and Somerville, C. 1993. Isolating plant genes. Trends Biotechnol. 11: 306-313.

Goodhead, D. T. 1994. Initial events in the cellular effects of ionizing radiations: Clustered damage in DNA. Int. J. Radiat. Biol. 65: 7-17.

Gottschalk, W. 1983. Induced Mutations in Plant Breeding. Springer, Berlin. Gray, J., Picton, S., Shabbeer, J., Schuch, W. and Grierson, D. 1992. Molecular biology of fruit ripening and its manipulation with antisense genes. Plant Mol. Biol. 19: 69-87.

Grosovsky, A. J. and Little, J. B. 1983. Influence of confluent holding times tb UV light mutagenesis in human diploid fibroblasts. Mutat. Res. llO: 401-41.

Hofte, M., Desprez, T., Amselen, J. et al. 1993. An inventory of 1152 expressed sequence tags obtained by partial sequencing of cDNAs from *Arabidopsis thatiana*. Plant J. 4: 1051-1061.

Howard-Flanders, p., west, S. c., and Stasiak, A. J. 1984. Role of ReCA spiral filaments in genetic recombination. Nature (London) 309: 215-220.

Howard-Flanders, P., Rupp, W. D., Wilkins, B. M. and Cole, R. S. 1968. DNA replication and recombination after UV-irradiation. Cold Spring Harbour Symp. Quant. Biol. 33: 195-205.

International Atomic Energy Agency, 1977. Manual on Mutation Breeding, 2nd edition, Technical Reports Series No. 119, IAEA, Vienna.

Jalil Miah, A. and Yamaguchi, H. 1964. Experiments on the induction of polygenic mutations with successive irradiation in rice. YTON 21 (2): 149-155.

Jalil Miah, A. and Yamaguchi, H. 1965. The variation of quantitative characters in the irradiated progenies of two rice varieties and their hybrids. Radiat. Botany 5: 187-196. Jarvis, P., Lister, C., Szabo, V. and Dean, C. 1994. Integration of CAPS markers into the RFLP map generated using recombinant inbred lines of *Arai1dopsis thaliana*. Plant Mol. Biol. 24: 685-687.

Jenner, T. J., Belli, M., Goodhead, D. T., Ianzini, F., Simone, G. and Tabocchini, M. A. 1992. Direct comparison of biological effectiveness of protons and alpha-particles of the same LET: III. Initial yield of DNA double-strand breaks in V79 cells. Int. J. Radiat. Biol. 61: 631-637.

Kallioniemi, A., Kallioniemi, O. P., Sudar, D., Rutovitz, D., Gray, J. W., Waldman, F. and Pinkel, D. 1992. Comparative genomic hybridization for molecular cytogenetic analysis of solid tumors. Science 258: 818–821.

Konieczny, A. and Ausubel, F. M. 1993. A procedure for mapping Arabidopsis mutations using co-dominant ecotype-specific PCR-based markers. Plant J. 4: 403-410.

Lawrence, C. W. 1965. Radiation-induced polygenic mutation. The use of induced mutations in plant breeding. Suppl. to Radiat. Botany 5: 491-496.

MacKey, J. 1954. Neutron and X-ray experiments in wheat and a revision of the speltoid problem. Hereditas 40: 65-180.

Maher, V. M., Dorney, D. J., Mendrala, A. L., Konze-Thomas, B. and McCormick, J. J. 1979. DNA excision repair processes in human cells can eliminate the cytotoxic and mutagenic consequences of ultra–violet radiation. Mutat. Res. 6: 311-323.

Mather, K. 1957. The Measurement of Linkage in Hereditv. Methuen and co.,

London. Mather, K. and Jinks, J. L. 1971. Biometrical Genetics, 2nd ed. Cornell Univ. Press, Ithaca.

Mortimer, R. K. 1958. Radiobiological and genetic studies on a polyploid series (haploid to hexaploid)of *Saccaromyces cerevisiae.* Radiat. Res. 9: 312-316.

Mourad, G., Haughn, G. and King, J. 1994. Intragenic recombination in the CSR1 locus of Arabldopsls. Mol. Gen. Genet. 243: 178-184.

Nagatomi, S., Miyahira, E. and Degi, K. 1996. Combined effect of gamma irradiation methods and *in vitro* explant sources on mutation induction of flower color in *hbyrsanthemum morifo1ium* Ramat. Gamma Field Symp. *35: 51-69.*

Nagatomi, S., Tanaka, A., Watanabe, H. and S. Tano. 1997. Chrysanthemum mutants regenerated from *in vitro* explants irradiated with $^{12}C^{5+}$ ion beam. Technical News, Inst. Radiat. Breed., No. 60.

Newman, T., de Bruijn, F. J., Green, P., *et al.* 1994. Genes galore: a summary of methods for accessing results from large-scale partial sequencing of anonymous *Arabidopsis* cDNA clones. Plant Phys. 106: 1241-1255.

Newman, H. C., Prlse, K. M., Folkard, M. and Mlchael, B. D. 1997. DNA double-strand break distributions in X-ray and alpha-particle irradiated V79 cells: Evidence for non-random breakage. Int. J. Radiat. Biol. 71: 347 -363.

Osone, K. 1963. Studies on the developmental mechanism of mutated cells induced in irradiated rice seeds. Japan. J. Breed. 13: 1-13.

Paterson, A. H. (ed)1998. Molecular Dissection of Complex Traits. CRC Press, New York.

Prise, K. M., Folkard, M., Davies, S. and Michael, B. D. 1990. The irradiation of V79 mammalian cells by protons with energies below 2 MeV: Part II. Measurement of oxygen enhancement ratios and DNA damage. Int. J. Rad1at. Biol. 58: 261-277.

Scossiroli, R. E. 1966. The use of induced genetic variability for quantitative traits after seed irradiation in *Triticum durum*. Savremen. Poljopr. 14: 221-234.

Sasaki, T., Song, J., Koga-Ban, Y. et al. 1994. Toward cataloguing all rice genes: large-scale squencing of randomly chosen rice cDNAs from a callus cDNA library. Plant J. 6: 615-624.

Sekiguchi, F., Yamakawa, K. and Yamaguchi, H. 1971. Radiation damage in shoot apical meristems of *Antirrhinum majus* and somatic mutations in regenerated buds. Radiat. Botany 11: 157-169.

Stadler, L. J. 1930. Some genetic effects of X-rays in plants. J. Heredity 21: 3-19.

Stone-Wolff, D. S. and Rossman, T. G. 1982. Demonstration of recovery from the potentially mutagenic effects of ultraviolet light by replication-inhibited Chinese hamster V79 cells. Mutat. Res. 95: 493-503.

Tanaka, A., Chantes, T., Shikazono, N., Yokota, Y., Watanabe, H. and Tano, S. 1996. Novel mutations induced by ion beams in *Arabidopsis thaliana*. JAERI TIARA Annual Report 1995, 32-33.

Tautz, D. 1989. Hypervariability of simple sequences as a general source for polymorphic DNA markers. Nucl. Acids Res. 17: 6463-6471. Uchimiya, H., Kidou, S., Shimazaki, T. et al. 1992. Random sequencing of cDNA libraries reveals a variety of expressed genes in cultured cells of rice (*Oryza sativa* L).. Plant J. 2: 1005-1009.

West, S. C., Cassuto, E. and Howard-Flanders, P. 1981. Mechanism of *E. coli* RecA protein-directed strand exchange in post-replication repair of DNA. Nature (London) 294: 659-662.

Williams, J. G. K., Kubelik, A. R., Livak, K. J., Rafalski, A. and Tingey, S. V. 1990. DNA polymorphisms amplified by arbitrary promoters are useful as genetic markers. Nucl. Acids Res. 18: 6531-6536.

Yamaguchi, H. 1962. Chimaeric formation in an Xl panicle after irradiation of dormant rice seed. Radiat. Botany 2: 71-77.

Yamaguchi, H. 1969a. Onset of DNA synthesis after soaking of rice seed as disclosed by the induction of mutations. Radiat. Botany 9: 341-348.

Yamaguchi, H. 1969b. The germinating stage in rice seed that produces the greatest yield of radiation-induced mutations. Mutat. Res. 8: 655–657.

Yamaguchi, H. 1973. The production of mutations by ionizing radiations and chemical agents in relation to the duplication of chromosomes. Gamma Feld Symp. ll: 29-42.

Yamaguchi, H. 1974. Mutations with gamma irradiation on dormant seeds of a partially asynaptic strain in rice. Japan. J. Genet. 49: 81-85.

Yamaguchi, H. 1979. Enhancement of gamma-ray induced mutation in barley seeds by inhibitors of the unscheduled DNA synthesis. Proc. 6th Intern. Cong. Radiat. Res. 575-581.

Yamaguchi, H. 1991. Mutation: history, classification and theories. Mandel, A. K., Ganguli, K. K. and Banerjee, S. P. (eds). Advances in Plant Breed–ing, Vol. 1: 169-186.

Yamaguchi, H. 1994. Experimental mutagenesis: Introductory review. J. Komazawa Jr. Col. 22: 37-48.

Yamaguchi, H. and Jalil Miah, A. 1964. Mutational mosaicism induced by X-, –rays and ethyleneimine in rice. Radioisotopes 13: 472-476.

Yamaguchi, H. and Matsubayashi, I. 1973. Mutational response of rice seeds to ethyl metanesulfonate and busulfan in relation to chromosome duplication. Mutat. Res. 17: 191-197.

Yamaguchi, H., Tatara, A. and Naito, T. 1975. Unscheduled DNA synthesis induced in barley seeds by gamma-rays and 4-nitroquinoline I-oxide. Japan. J. Genet. 50: 307-318.

Yamaguchi, H., Tatara, A. and Naito, T. 1976. Accessibility of -ray induced primer toward DNA polymerase I of *Escherichia coli* during soaking of barley seed. Environ. Explt. Botany 16: 141-144.

Zamecnik, P. and Stephenson, M. 1978. Inhibition of Rous sarcoma virus replication and cell transformation by a specific oligodeoxynucleotide. Proc. Natl. Acad. Sci. U. S. A. 75: 280-284.

Chapter 2

Mutation Breeding for Crop Improvement: A Review

M.A. Awan

Nuclear Institute for Agriculture and Biology (NIAB), Faisalabad, Pakistan

INTRODUCTION

The discovery that ionizing radiation and chemical mutagens could cause genetic changes in an organism and modify linkages offered promise in the improvement of crop plants. Mutation breeding involves the use of induced beneficial changes/mutations for practical plant breeding purpose both directly as well as indirectly. Mutation breeding can be used to complement and supplement existing germplasm resources (Konzak *et al.*, 1976). Mutagenic treatment can give rise to i} new alleles of known genes and alleles of previously unknown genes, ii} create useful genetic variability in modern phenotype, iii} modify likages and retain desirable complexes, iv} create isogenic genetic variants for biochemical, genetic and physiological analysis. Induced mutations are now widely used for introducing genetic changes, creation of new genetic resources and breakage of linkages. Although the occurrence of mutation is a random process and the probability of getting mutation is very low (a particular gene can be expected to mutate once in 10,000 mutagen treated cells provided that the effective treatment was given (Yonezawa and Yamazata1977, Brock1979); yet it can be increased if done under correctly estimated conditions.

Organic evolution has proceeded through small changes occurring in the organisms. The improved changes were retained through natural selection. Mutation breeding is a similar procedure with the only difference that efforts are made to induce greater changes and selections are made under more rigorous conditions and thus the process is hastened. As a result of natural mutations and other factors, the organisms harmful to agricultural crops are also changing and adapting to new situations. The plant breeding programmes have to keep pace with these changes since a variety which is resistant to one race/biotype of disease/insect may not be resistant to the new race/biotype (there may be other reasons for breakdown of resistance as well). By use of induced mutations shortage of useful germplasm resistant to diseases/insects can also be overcome.

Historical

First observations about artificial induction of genetic changes date back to the beginning of the 20th century (Gager 1908), but proper proof of Mendelian inheritance of such induced changes came only in the late twenties by Muller, Stadler and others using X-rays as mutagen (Muller 1927, Stadler 1928a, b). Although Muller, being an entomologist, assumed that induced mutations could play an important role in future genetic improvement of plants, Stadler as a plant breeder became rather sceptical about such prospects when he noticed so may useless and even deleterious mutations in maize and barley. Stadler's specticism has influenced almost two generations of plant breeders, especially in North America, and has led to a widely spread preconceived notion that mutation induction will be of high interest to geneticists, but is a rather wasteful undertaking for plant breeders. Stadler's view, primarily, was based upon his experiments with maize, where a lot of genetic diversity exists (as in most cross-pollinated crops), from which improved varieties could still easily be developed simply through selection or a combination of cross breeding and selection.

Among the first researchers who used mutagenesis strictly for plant breeding were Freisleben and Lein in Halle (Germany). They succeeded in obtaining mildew resistance in barley (Freisleben and Lein 1942) and developed a practical mutation breeding procedure (Freisleben and Lein 1943a, b), but due to World War II this work was not followed up properly (Hoffmann 1959). In the meantime,primarily in Sweden, plant geneticitst such as Nilsson-Ehle, Gustafsson, Hagberg, Gelin and Nybom continued to experiment mainly with X-rays and carried out rather systematic studies as to optimal doses, treatment conditions, mutation frequency and mutation spectra. They also compared X-ray effects with those of certain chemicals which became known as mutagens, such as (EI) ethylenimine (Micke *et al.*, 1987). Although most of this work was of a fundamental nature, there were by–products which turned out to be of interest to breeders–easily recognizable mutants of barley, wheat, oats with early or late *r* heading, short straw or different spike architecture but also mutants of pea, soybean, flax, mustard and rape (Gustafsson 1947, Mac Key 1956).

By 1950 mutation induction research began to flourish in several other countries such as USA, Italy, France, USSR, Netherlands and Japan, using mainly 60Co gamma rays or neutrons obtainable from various newly established nuclear research centres (Scarascia Mugnozza 1966). For more than 10 years, major research efforts went into the search for radiation treatment conditions or additional treatments (before or after irradiation) that could modify the random mutation induction into something more specific, more directed, more economically useful (Nilan *et al.*, 1965). Water, oxygen and time were the main factors discovered to be of influence, but their deliberate control only brought about quantitative differences, which could also be obtained from different doses, and did not really lead to any useful methodological improvement (IAEA 1961, 1965). Later on, developing countries began to play an increasing role in mutation breedingwork. Particularly in Asia. New varieties of rice soon appeared on the market which derived some valuable characteristics from mutation induction (IAEA 1971, Sigurbjornsson and Micke 1969, 1974, Wang 1986). In the beginning, mutation breeding was based primarily upon x-rays, but now mainly gamma rays and to a smaller extent fast or thermal neutrons also started to be used.

In 1969, the Joint FAO/IAEA Division started to organize course for plant breeders on the induction and use of mutation, and in the same year published the first edition of the Manual and Mutation Breeding. It may therefore be justified to consider 1969 as the year that marked the establishment of mutation breeding as a practical tool available to plant breeders in their endeavours to develop more productive cultivars with better resistance to stresses, pathogens and pests, and with improved quality characteristics for plant products used as food, feed or industrial raw material.

Mutation Spectrum

Known mutant collections, of course, contain only selected mutants, mostly of easily recongnizable type and therefore not fully representative of the potential spectrum of induced mutations. A specific advantage of mutation induction, however, is the possibility of obtaining unselected genetic variation, whereas all other available germplasm has already passed screens of selection by nature or man. The question whether induced mutations duplicate the genetic variation produced by nature (e.g. Allard 1960, Herskowitz 1962) is rather theoretical since both natural or man-made germplasm do not represent all the possible spontaneous mutations or recombinants. When new breeding objective come up–and this will be more often in the future–it will be a matter of lucky chance if the desired variant exists among stocks in germplasm collections or in habitats of high diversity. Spontaneous mutation rates, on the other hand, will not give much new variation to breeders. Today it seems somewhat strange that early mutation researchers were disappointed about the randomness that gives mutation induction a unique role among r', the plant breeder's tools. There is sufficient evidence that induced mutations fit Vavilov's law of homologous genetic variation (Scholz and Lehmann 1958, Enken 1967 Stubbe 1967). It has logically been concluded that limitations of mutation breeding are not in mutagenesis as such but rather in identification and selection of desired variants (Gregory 1956, IAEA 1984a).

Achievements

An early record of an induced valuable mutant has been shown by Ramiah and Rao (1953). They reported about 36 X-ray induced mutations affecting different characters in rice. Of these one mutant proved useful from the economic point of view. It had a slight shorter stature with a large number of tillers than the original parent material and proved valuable in that it performed well in rich soil where problem of lodging was serious. Looking at the progress of mutation breeding, it seems that as far as cereals are concerned major emphasis has been on obtaining mutants for improved disease resistance and improved grain quality (protein), but main results were in improving lodging resistance (short or/and stiffculm) (IAEA 1984c, Maluszynski *et al.*, 1986) and altering crop duration *i.e.* photoperiod sensitivity (Awan *et al.*, 1982, Gottschalk and Wolff 1983, Donini *et al.*, 1984, Konza 1984). Results in terms of improved grain protein were not discouraging, but remained below the rather exaggerated expectations (Micke 1983a, IAEA 1984b, Muller 1984, Awan & Cheema 1988). This on one hand, is certainly due to the low heritability of quantitative endosperm characters and, r consequently, the inefficient selection. With regard to disease resistance, applied selection procedures generally have been inadequate to a large extent because objectives were poorly defined due to insufficient understanding of epidemiological principles and host/parasite interaction. Nevertheless, some results have been rather spectacular (IAEA 1977b, 1983a, Konza 1984). On the other hand, it is worth noting that more than 40 years after its discovery one has begun to understand the nature of mutations in the famous ml-o locus of barley and the reasons for the universal, non-specific resistance rendered by a series of recessive alleles in that locus (Jorgensen 1975, Sokou 1982, Shou *et al.*, 1984). It is also the barley powdery-mildew complex where first clear experimental proof was obtained as to the possibility of improving quantitative resistance by monogenic mutations (Robbelen Abdel-Hafez and Reinhold 1977, Abdel-Hafez and Robbelen 1979, 1981, Aziz *et al.*, 1980).

Since most mutation breeding work was performed with annual and self-pollinating cereals, most experiences relate to them. The problems in other groups of crop plants, however, are quite different. For example, in grain legumes, where breeding advances lag far behind the cereals, we have still a relatively poor adaptation of the plant architecture to modern farming systems. The plant architecture, of course, being the ultimate result of numerous physiological reactions and interactions, is therefore not likely to be inherited as simply as the culm length in cereals (Micke 1979, 1984a). On the other hand, reports confirm that even with single monogenic mutations a remarkable r reconstruction of plant architecture is achievable in grain legumes and in other dicotyledonous plants, *e.g.* in chickpea (Shaikh *et al.*, 1980), pigeon pea and mungbean (Rao *et al.*, 1975), pea (Jaranowski and Micke 1985), castor bean (Kulkarni 1969), cotton (Raut *et al.*, 1971 Swaminathan 1972), linseed (George and Nayar 1973 Nayar 1974). Fast development of computer technology enabled FAO/IAEA to organize the data base in 1987. The information contained in this data base is based on data on mutant cultivars published in various issues of Mutation Breeding Newsletter. According to the latest information available there are 1239 accessions in the FAO/IAEA Mutant Varieties Database (Maluszynski *et al.*, 1995). These crop varieties were developed either directly after mutagenic treatment or through crosses involving mutant varieties or mutant lines. The cumulative number of officially released mutant cultivars indicates that more than 50 per cent of these varieties were released during the period between 1980-1995. Maximum number of crop varieties 304 have been released in China followed by India (243), the former USSR and the Russia Federation (209), the Netherlands (176), Japan (115) and USA (93). Mutant cultivars of cereal dominate (828) followed by legumes, oil crops, and industrial crops. In cereals mutation techniques were most successfully applied for improving rice (322 mutant cultivars) and barley (240) followed by wheat, maize, durum wheat and other cereals such as oats, millet, pearl millet etc. Application of mutation techniques for improving a particular crop or a group of crops has been the subject of review papers published by the International Atomic Energy in Mutation Breeding Reviews (Hanna 1982, Jaranowski and Micke 1985, Daskalov 1986, Spiegel-Roy 1990, Robbelen 1990, Rutger 1992, Micke *et al.*, 1993, Scarascia Mugnozza *et al.*, 1993).

Achievements Made in Pakistan

Realising the potential role of induced mutations for the improvement of crop plants, the Pakistan Atomic Energy Commission initiated a modest research activity in 1963 at the Atomic Energy Agricultural Research Centre, Tandojam. Two other agricultural research institutes, Nuclear Institute for Agriculture and Biology (NIAB), Faisalabad and Nuclear Institute for Food and Agriculture (NIFA), Peshawar were established in 1972 and 1982 respectively. These institutions are now involved with a wide variety of biological and agricultural research using nuclear and other advanced techniques. One of their important activities is the crop improvement programme which supplements the efforts of the conventional breeders where induction of mutation provides a better or the only approach to evolving the desired genotype.

The efforts of the three institutions has resulted in evolving 29 improved varieties of wheat, rice, chickpea, mungbean and cotton (Table 2.1). These crops were selected because of little variability available in them to improvement through conventional techniques. The main objectives have been to confer specific changes such as improvement of plant structure, reduction of growth duration of the crop, resistance against diseases and pests, physiological characters *i.e.* heat tolerance, uniform maturity, photoperiod insensitivity etc., in the native well adapted crop varieties to make them more productive. Some of the salient results are described here.

Wheat

From the variability created by irradiation of seed with the predetermined doses of gamm rays, mutants were selected for improved traits such as stiff stem, early maturity, higher protein content and disease resistance etc. Eight wheat varieties namely Jauhar 78, Sind 81, Sarabz, Saughat-90, Bakhtawar-92, Kiran-95 and Tatar has been released as commercial varieties for general cultivation.

Rice

In mutation breeding programme of rice three types of cultivars has been used i) the local non aromatic variety Kangni-27, ii) the local aromatic varieties–Basmati types and iii) IRRI type varieties. Early maturity and short stature have been the most frequently selected features in these materials.

From IRRI varieties mutants fine grain 6 (NIAB-6) AND IR-8-5 were selected. Mutant IR-6-18 possess long and fine grain with higher grain yield and has been approved as a commercial variety Shadab. Mutant FG-6 (NIAB-6) is a derivative of IR-6. It has long and transluscent grain. NIAB-6 is also salt tolerant and has been proposed for approval as a commercial variety for saline environments. Another salt tolerant mutant has been released as a commercial variety 'Shua' at AEARC, Tandojam. Mutations were

Table 2.1: Crop Varieties Developed Through the Use of Induced Mutations at Agricultural Research Institutes in Pakistan

Crop	*Variety*	*Year of Release*	*Institution*
Cotton	NIAB-78	1983	NIAB
	NIAB-88	1990	NIAB
	NIAB-26	1992	NIAB
	CHANDI	1995	AEARC
	NIAB-KARISHMA	1996	NIAB
Rice	Kashmir Basmati	1977	NIAB
	Shadab	1987	AEARC
	Shua	1992	AEARC
	Khushboo	1995	AEARC
Wheat	Jauhar-78	1979	AEARC
	Sind-81	1982	AEARC
	Sarsabz	1986	AEARC
	Saughat-90	1990	AEARC
	Bakhtawar	1992	NIFA
	Kirin	1985	AEARC
	Tatara	1985	NIFA
	Nishtar	1995	NIFA
Chickpea	CM-72	1983	NIAB
	NIFA-88	1990	NIFA
	CM-88	1995	NIAB
Mungbean	NIAB-Mung-28	1983	NIAB
	NIAB-Mung-121-25	1986	NIAB
	NIAB-Mung-19-19	1986	NIAB
	NIAB-Mung-20-21	1986	NIAB
	NIAB-Mung-13-1	1986	NIAB
	NIAB-Mung-51	1990	NIAB
	NIAB-Mung-54	1990	NIAB
	NIAB-Mung-92	1996	NIAB
Rapeseed	Abasin	1995	NIFA

induced for reduction in height and growth duration in the tall and long duration variety Basmati 370. Mutant EF-29-1 was selected for ealry maturity (100 days). It proved ideal for northern hilly areas of the country where the summer is short and Basmati 370 does not get to maturity because of early onset of winter. On the basis of its early maturity and cold tolerance, the mutant was approved as a commercial variety 'Kashmir Basmati', and is being cultivated in those areas. Several semi–dwarf mutants were selected from basmati varieties and used in genetic studies (Awan *et al.*, 1986). These mutants also posses better lodging resistance and high yield potential and quality parameters at par with Basmati 370. Of them DM-25 has been porposed for release as a commercial variety. An aromatic mutant variety I Khushboo' derivative of Jajai-77 has also been released for general cultivation in Sind.

Grain Legumes

Efforts were made to induce mutations to improve the plant induce resistance against diseases in Mungbean In mungbean, several mutants with reduced height, uniform and ealry maturity of pods and higher grain yield were isolated. Of these Mutant M-28 which matures in 80 days as against 90 days of parent was released as variety NIAB Mung 28 in 1983. Four more mutants, NIAB Mung 19-19 and 121-25 maturing in 60-70 days, and NIAB Mung 20-21 and 13-1, maturing in 55-58 days, yielding about 30 per cent higher than the parent types and maturing uniformly, were released as commercial varieties in 1986. The crosses between local small-seeded variety 6601 and *r* exotic large seeded variety VC 1973A followed by irradiation of hybrid seed led to the development of several large seeded, high yielding and disease resistant lines which thrive both in spring and summer. Of these two lines NM 51 and NM 54 were released as commercial varieties namely NIAB Mung 51 and NIAB Mung 54 in 1990. Another high yielding large seeded line NM 92 was approved as a commercial variety NIAB Mung 92 for general cultivation in 1996.

In case of chickpea, mutations were induced mainly for resistance against blight disease which caused havoc in the existing cultivars in early eighties. Induction of mutations have led to creation of a wide variability in chickpea. Selection for blight resistance and trials at farmers'field resulted in the evolution of a blight resistant variety CM-72 introduced for commercial cultivation in 1983. Eversince, the chickpea production has been stabilized. Another mutant line CM-88, tolerant to both blight and wilt, obtained from a different genetic background has been released a variety 'CM 88' for general cultivation in 1995. Two blight resistant mutants were released as commercial varieties namely NIFA 88 and NIFA 95 during 1990 and 1996 respectively for general cultivation.

Cotton

In cotton, better results were obtained when F1 hybrid seed was irradiated. Using several local (AC 134, B 557 and exotic (Deltapine, Stoneville etc). cultivars, crosses were made and F1 hybrid were irradiated. From the variability thus created, selections were made for erect monopodial plant with semi hairy *r* leaf and shorter duration.From among the selections, line NIAB-78 had the desired fibre quality and gave the highest yield in multilocational trials. It was released as a commercial variety during 1983. This variety has proved to be adapted to the different agroclimatic areas of the country. Since its introduction the cotton production registered a quantum jump in Pakistan (Saeed Iqbal *et al.*, 1991). NIAB-78 was crossed with an insect tolerant line and evolved a variety NIAB-86 which is comparatively more suited to salt affected and lower fertility areas. It was released as commercial variety in 1990. Crossing mutant line with exotic germplasm having nectariless leaf resulted in the evolution of NIAB-26 which was released in 1992. An other nectariless line that had tolerance to heat and leaf curl virus disease was released as a commercial variety I Karishma' in 1996. At AEARC, Tandojam a high yielding mutant variety was released as commercial variety 'Chandi' in 1996. A high yielding variety of Rape seed namely 'Abasin 95' has been released for commercial cultivation in 1996.

From the foregoing, it can be concluded that the use of induced mutations for crop improvement has had to the development of several improved varieties of crops in several countries which have played significant role in increasing agricultural production the world over. In Pakistan 29 improved varieties of different crops have been developed which clearly indicates the potential of this technique. In addition a wealth of genetic variability has been developed for use in the cross breeding programmes and a few varieties of cotton have been developed in Pakistan by using an induced mutants as one of the parents. However, the method is to be used selectively for crops where wider variability is desired or where conventional technique presents problems. Further that there are certain characters more amenable to change; crops in which such characters are desired to be changed need to be chosen for improvement.

REFERENCES

Abdel-Hafez, A.A.G.I. and Robbelen, G. 1979. Differences in partial resistance of barley to powdery mildew (*Erysiphe graminis* DC. F. sp. hordei Marchal) after chemomutageness. I. Screening of mutants under field conditions, *Z.Pflanzenzucht*. 83: 321-339.

Abdel-Hafez, A.A.G.I. and Robbelen, G. 1979. Differences in partial resistance of barley to powdery mildew (*Erysiphe graminis* DC. F. sp. hordei Marchal) after chemomutageness. III. Agronomic performance of the mutants, *Z.Pflanzenzucht*. 86: 99-109.

Allard, R.W. 1960. Principles of Plant Breeding, New York, London, John Wiley and Sons Inc.

Awan, M.A., Maqbool Ahmed, and Cheema, A.A. 1982. Evaluation of short stature mutants of Basmati 370 for yield and grain quality characteristics. Pak. J.Sci. Ind. Res. 25(3): 67-71.

Awan, M.A., Cheema, A.A., and Tahir, G.R. 1986. Induced mutations for genetic analysis in rice. in: Rice Genetics, IRRI, Philippines, pp 697-705.

Awan, M.A., and Cheema, A.A. 1988. New mutant genes for early maturity and dwarfism in Basmati rice (*Oryza sativa* L). SABRAO J. 20(5): 56-61.

Aziz, A.Abdel-Hafez, G.I. and Robbelen, G. 1980. Differences in partial resistance of barley to powdery mildew (*Erysiphe graminis* DC f. sp. hordei Marchal) after chemomutagenesis. II. Reaction of mutants to pathotypes, *Euphytica* 29:755-768.

Brock, R.D. and Micke, A. 1979. Economic aspects of Mutations for Crop Improvement in Africa, Vienna, International Atomic Energy Agency, TEC-DOC-222 pp 19-32.

Daskalov, S. 1986. Mutation breeding in pepper, Mutat. Breed. Rev. 4 1-26.

Donini, B., Kawai, T. and Micke, A. 1984. Spectrum of mutant characters utilized in developing improved cultivars, in: Selection in Mutation Breeding Vienna, International Atomic E~rgy Agency, pp 7-31.

Enken, V.B. 1967. Manifestation of Vavilov's law of homologous series in hereditary variability in experimental mutagenesis, in: Induced Mutations and their Utilization, Berlin, Akademie-Verlag, pp 123-129.

Freisleben, R. and Lein, A. 1942. Uber die Auffindung einer mehltauresistenten MUtante nach rontgenbestrahlung einer anfalligen Linie von Sommergerste, Naturwissenschaften 30: 608.

Freisleben, R. and Lein, A. 1943a. Vorabeiten zur zuchterischen Auswertung rontgeninduzierter Mutationen I, Z. Pflanzenzucht. 25: 235-254.

Freisleben, R. and Lein, A. 1943b. Vorabeiten zur zuchterischen Auswertung rontgeninduzierter Mutationen II, Z. Pflanzenzucht. 25: 255-283.

Gager, C.S. 1908. Effects of the rays of radium on plants, Mem. N.Y.Bot. Gard. 4: 278.

George, K.P. and Nayar, G.G. 1973. Early-dwarf mutant in inseed induced by gamma-rays, Curro Sci. 42: 137-138.

Gottschalk, W. and Wolff, G. 1983. Induced Mutations in Plant Breeding. Monographs on Theoretical and Applied Genetics No.7, Berlin, Springer-Verlag.

Gregory, W.C. 1956. Induction of useful mutations in the peanut, Brookhaven Symp. Biol. 9: 177-190.

Gustafsson, A. 1947. Mutation in agricultural plants. Hereditas 33: 1-100.

Hanna, W.W. 1982. Mutation breeding in pearl millet and sorghum, Mutat. Breed. Rev. 1: 1-13.

Herskowtz, I.H. 1962. Genetics, Boston, Toronto, Little, Brown and Company.

Hoffmann, W. 1959. Neuere Moglichkeiten der Mutationszuchtung Z. Pflanzenzucht. 41: 371-394

IAEA 1961. Effects of ionizing Radiations on Seeds. (Proceedings of a symposium) Vienna, International Atomic Energy Agency.

IAEA 1965. The use of Induced Mutations in Plant Breeding. (Report of the FAO/IAEA Technical Meeting. Rome, 1964), Oxford, Pergamon Press.

IAEA 1971 Rice Breeding with Induced Mutations III., Vienna, International Atomic Energy Agency.

IAEA 1977b. Induced Mutations Against Plant Diseases, Vienna, International Atomic Energy Agency.

IAEA 1983a. Induced Mutations for Disease Resistance in Crop Plants II, Vienna, International Atomic Energy Agency.

IAEA 1984a. Conclusions and recommendations, in: Selection in Mutation Breeding, Vienna, International Atomic Energy Agency, 157-169.

IAEA 1984b. Cereal Grain Protein Improvement, Vienna, International Atomic Energy Agency.

IAEA 1984c. Semidwarf Cereal Mutants and their Use in Cross–Breeding II, Vienna, International Atomic Energy Agency, TEC-DOC-307.

Iqbal, R.M.S., Chaudhry, M.B., Aslam, M., and Bandesha, A.A. 1991. Economic and Agricultural Impact of Mutation Breeding in Cotton in Pakistan–A review, in: Plant Mutation Breeding for Crop Improvement, Vienna, International Atomic Energy Agency. pp 187-201.

Jaranowski, J. and Micke, A. 1985. Mutation breeding in peas, Mutation Breeding Review No.2, Vienna, International Atomic Energy Agency.

Joergenson, J.H. 1975. Identification of powdery mildew resistant barley mutants and their allelic relationship, Barley Genetics III, 446-455.

Konzak, C.F., Nilan, R.A., and Kleinhofs, A. 1976. Artificial mutagenesis as an aid in overcoming genetic vulnerability of crop plants, in: Genetic Diversity in Plants, (Eds Muhammad Amir., Askel, Rustem and Von Borstel, R.c). Plenum Press, New York, pp 163-177.

Konzak, C.F. 1984. Role of induced mutations, in: Crop Breeding, a Contemporary Basis, (Eds Vose, P.B. and Blixt, S.G). Oxford, Pergamon Press, pp 216-292.

Kulkarni, L.G. 1969. Induction of useful mutations in castor. Radiations and Radionimetic Substances in Mutation Breeding, Proceedings of a Symposium, Bombay, pp 293-299.

Maluzynski, M., Micke, A. and Donini, B. 1986. Genes for semi–dwarfism in rice induced by mutagenesis, in: Rice Genetics (Proc. International Rice Genetics Symposium, Los Banos, Philippines, 27-31 May 1985), Manila, International Rice Research Institute, pp. 729-737.

Malusznski, M., Vanzanten, L., Ashri, A., Brunner, H., Ahloowalia, B., Zapata, F.J., and Weck, E. 1995. Mutatioin techniques in Plant Breeding, in: Induced Mutations and M<11ceular Techniques for Crop Improvement, International Atomic Energy Agency. pp. 489-504.

Micke, A. 1979. Use of mutation induction to alter the ontogenic pattern of crop plants, in: Crop Improvement by Induced Mutation, Gamma Field Symposia No.18, Ohmiya, Japan, Institute of Radiation Breeding, pp 1-23.

Micke, A. 1983a. International research programmes for the genetic improvement of grain proteins, in: Seed Proteins: Biochemistry, Genetics, Nutritive Value, (Eds Gottschalk, W. and Muller, H.P). The Hague, Boston, London, Martinus Nijhoff/Dr. W.Junk Publishers, pp 25-44.

Micke, A. 1984a. Mutation breeding of grain legumes, Plant Soil 82: 337-358.

Micke, A., Donini, B., Maluszynski, M. 1993. Les mutations induites en amelioration des plantes, Mutat. Breed. Rev. 9: 1-44.

Muller, H.J. 1927. Artificial transmutation of the gene, Science 66: 84-87.

Muller, H.P. 1984. Breeding for enhanced protein, in: Crop Breeding, a Contemporary Basis, Oxford, Pergamon Press, pp 382-399.

Nayar, G.G. 1974. Yield potential of a radiation induced early–dwarf mutant in linseed, in: Use of Radiations and Radioisotopes in Studies of Plant Productivity. Proceedings of a Symposium, pantnagar, India, pp 109-117.

Nilan, R.A. Konzak, C.F., Wagner, J. and Legault, R.R. 1965. Effectiveness and efficiency of radiations for induced genetic and cytogenetic changes, in: the Use of Induced Mutations in Plant Breeding, (Report of the FAO/IAEA Technical Meeting, Rome 1964), Oxford, Pergamon Press, pp 71-89.

Ramiah, K. and Rao, M.B.V.N. 1953. Rice breeding and genetics. Indian C.Agri. 40: 335-355.

Rao, C.H., Tickoo, J.L., Ram, H. and Jain, H.K. 1975. Improvement of pulse crops through induced mutations: Reconstruction of pant type, in: Breeding for Seed Protein Improvement using Nuclear Techniques, Vienna, International Atomic Energy Agency, pp 125-131.

Raut, R.N., Jain, H.K. and Panwar, R.S. 1971. Radiation induced photo-insensitive mutants in cotton, Curre. Sci. 40 383-384.

Robbelon, G. Abdel-Hafez, A.F., and Reinhold, M. 1977. Use of mutants to study host/pathogen relations, in: Induced Mutations against Plant Diseases, Vienna, International Atomic Energy Agency, pp 359-374.

Robbelen, G. 1990. Mutation breeding for quality improvement–A case study for oil-seed crops. Mutat. Breed. Rev. 6, 1-44.

Rutger, J.N. 1992. Impact of mutation breeding in rice–A review, Mutat. Breed. Rev. 8, 1-24. 17 'O}

Scarascia-Mugnozza, G.T. 1969. MUtazioni indotte e miglioramento genetico delle piante agarie, Genet. Agrar. 20: 140-178.

Scarascua-Mugnozza, G.T. *et al.*, 1993. Mutation breeding for durum wheat *(Triticum turgidum* ssp. *durum* Desf). improvement in Italy, Mutat. Breed. Rev. 10: 1-28.

Scholz, F. and Lehmann Ch. 1958. Die Gaterslebener Mutanten der Saatgerste in Baziehung zur Formen-mannigfaltigkeit der Art HOrdeum vulgare L.S. 1.1., Kulturpflanze 6: 123-166.

Shaikh, M.A.Q., Ahmed, Z.U. Majid, M.A.Bhuiya, A.D., Kaul, A.K. and Miz, M.M. 1980. Developent of a high yielding chickpea mutant, Mutation Breeding Newsletter NO.16 1-3, IAEA, Vienna.

Sigurbjornsson, B. and Micke, A. 1974. Philosophy and Accomplishments of Mutation Breeding. in: Polyploidy and

Induced MUtations in Plant Breeding, Vienna, International Atomic Energy Agency, 303-343.

Skou, J.P. 1982. Callose formation responsible for the powdery mildew resistance in barley with genes in the ml-o-locus, Phytopathol. Z., 104: 90-95.

Skou, J.P., Joergensen, J.H. and Lilholt, U. 1984. Comparative studies on callose formation in powdery mildew compatible and incompatible barley, Phytopathol Z. 109: 147-68.

Spiegel-Roy, P. 1990. Economic and agricultural impact of mutation breeding in fruit trees, Mutat. Breed. Rev. 6: 1-44.

Stadler, L.J. 1928a. Mutations in barley induced by X-rays and radium, Science 68: 186-187.

Stadler, L.J., 1928b. Genetic effects of X-rays on maize, Proc. Natl Acad. Sci. 14: 69-75.

Swaminathan, M.S. 1972. Mutational reconstruction of corp ideotypes, in: Induced Mutations and Plant Improvement, Vienna, International Atomic Energy Agency, pp 155-171.

Wang, L., Fan, Q., Shi, J. and Wang, Z. 1986. Studies on improving efficiency for inducing mutation of wheat hybrid by irradiation, in: Proc. International Symposium on Plant Breeding by Inducing Mutation and In-Vitro Biotechniques, Beijing (China), 16-20 October 1985, pp 39-44.

Yonesawa, K. and Yamagata, H. 1977. On the optimum mutation rate and optimum dose for practical mutation breeding, Euphytica 26: 413-426.

Chapter 3

Mutation Breeding in *Nigella sativa* L. (Black Cumin)

A.K. Biswas

Cytogenetics and Plant Breeding Laboratory, Department of Botany, University of Kalyani, Kalyani, (West Bengal), India

ABSTRACT

Nigella sativa L., the black cumin, is an important spice yielding plant. Economically it is important not only as a food flavouring agent but also for its therapeutic properties. Reports of treatment with mutagenic agents in *Nigella sativa* L. were available only in late seventies following x-irradiation. Mostly cytological anomalies were stated in the earlier observations.

Comprehensive studies on induced mutagenesis in *Nigella sativa* L. and *N. damascena* L. were carried out in the present author's laboratory following treatments with EMS (ethyl methane sylphonate) x-ray and gamma-radiations. The observed mutagenic consequences were assessed cytogenetically at the initial stage. Subsequently 25 types of mutations including 12 different chlorophyll mutations were identified. Four of the chlorophyll mutations namely *chloroxantha, marginata, coerulovirens* and *viridis* were viable, but only *viridis* could be evaluated in the advanced generations. Nine of the 13 other categories of mutations, namely *lax branching, feathery leaf, bushy, male sterile, crumpled leaf, early flowering, brown seed coat, dwarf* and *prostrate* were viable. The mutants were screened cytogenetically by analysing meiotic consequences, pollen and seed sterility and modifications in several other parameters mostly related to yield. All of these mutants were true breeding. Barring *viridis* and *chloroxantha* (showing digenic control), all other mutants showed monogenic inheritance. Mutagenic effectiveness and efficiency of the employed doses of the different mutagenic agents were estimated. Seven of these mutants were characterised by evaluation of seed protein content, protein profiles, protease and amylase activity and isozyme patterns. General and specific combining ability (gca and sca) were estimated in the mutants *early flowering, lax branching, feathery leaf* and *viridis* by diallel analysis. Gca analysis revealed that *lax branching* mutant was superior to others for transmission of capsule number/plant and number of secondary branches, *early flowering* mutant was the best parental genotype for transmission of earliness in flowering as well as maturity and tallness. Pureline selection would be the effective means for direct improvement of the traits concerned. Sca estimation suggests that *early flowering* × *lax* branching is the best combination for transmission of height and maternal influence plays significant role in it, highest number of capsule/plant is also transmitted through this cross combination. Transmission of earliness was best effected through the combination *Suttons local* variety × *early flowering*. *Lax branching* × *early flowering* was also good specific combiner for transmission of earliness. The cross combinations *lax branching* × *feathery leaf and early flowering* × *viridis* were best for transmission of 100 seed weight and total yield/plant respectively. Both additive and non-additive genes are involved in the expression of all the traits but primary branches/plant which is controlled by additive genes only. The parental genotypes and cross combinations thus identified from induced mutant progenies offer scope of improvement through proper selection in black cumin.

INTRODUCTION

Nigella sativa L., commonly known as black cumin or small fennel, is an important spice yielding crop plant. Use of this plant dates back to Hippocrates, who described black cumin seed as a tonic condiment (vide Watt 1972). It is known to be a native of Levant and cultivated or occasionally found as a weed in Punjab, Himachal Pradesh, Bihar and Assam (Anonymous 1991). In India, particularly in West Bengal, and Bangladesh it is cultivated as a winter crop for its seed being consumed as a food flavouring agent. The genus *Nigella* belongs to the family Ranunculaceae and comprises of 20 species (Willis 1967). Four of which are cultivated. Among these, *Nigella sativa* L. is of adequate economic significance for its aromatic seeds known to be black cumin. *N. damacena* L, a close relative of black cumin, is grown for its showy flower with double whorled petaloid calyx. The other species are not cultivated in India.

N. sativa L. is an annual, diffusedly branched herb. Leaves are alternate with highly incised blades and sheathing base. Flower is solitary, bisexual, hypogynous with petaloid sepals, petals are absent. Stamens are numerous, free and spirally arranged on the receptacle. Carpels are usually 5 (3-8), ovary chambers are as many as number of carpels, style small, stigma free, elongated, 3-8. Flowers are protandrous, anthers on maturity of stamen bend outwards from the centre. Flowers are self-pollinated stigma on maturity bend outwards to touch the mature authers when pollen grain are available. Fruit is dry, dehiscent, capsule, seeds are many, ellipsoid, triangular and blackish in colour with rugose surface.

Seeds of black cumin are used as flavouring agent in the preparation of curries vegetables etc. These are also used in bakeries and confectioneries due to their pungency and aromatic taste. Use of black cumin seeds for relief in cold cough and bronchial asthma is prevalent. For aromatic, carminative stomachic and digestive properties the seeds are recommended as effective remedial agents for indigestion, loss of apetite, diarrhoea etc. They are also used popularly as aromatic adjuncts to purgative or bitter remedies. Anticancerous activity of alcoholic extracts of seed was reported (Abdel *et al.*, 1992).

Diethyl ether extracts of black cumin seeds have antibacterial activity (Hanafy and Hatem 1991). Hypoglycemic effect of volatile oil against diabetes and blood pressure was mentioned (Al-Hader *et al.*, 1993). Inhibition of Jarkat T-cell leukemia by volatile oil was demonstrated (Hailat *et al.*, 1995).

N. sativa L. is not cultivated properly on any considerable scale in spite of its commercial importance as spice and manifold uses. Seeds are collected mostly from plants growing wild in forest areas for use as flavouring material and for medicinal purpose (Anonymous 1991). The main constraints of its cultivation and yield productivity are possibly its sensitivity to climatic fluctuation and shattering of fruits resulting in loss of seed yield. It is grown once in a year during rabi season in any good soil from October to November in plains and from April to May in hills (Chakraborty and Chakraborty 1964). Compared to other rabi crops yield of black cumin is not satisfactory and can not keep pace with the necessity of population of our country although it possesses considerable economic significance and has the potentiality to be grown as a rotational crop.

The plant species is characterised by its solitary inflorescence having only a single flower per branch. Yield per plant may be enhanced provided number of branches, number of fertile flowers and capsules per branch, size and number of compartments of fruits and seed weight increase. It is self pollinated and exhibits only a limited range of variations offering rather little scope of improvement through conventional breeding methods. New variations may, however, be generated through induction of mutation within a rather short period of time. Sigurbjornsson and Micke (1974) considered mutation breeding as a valuable supplement to conventional breeding methods. Most important aspects of mutation breeding are induction, identification, isolation and use of drastic changes of phenotype brought about by mutational events among the genes (Scossiroli 1965). Recombination and variability may be achieved by breaking gene linkages through treatment with mutagens, particularly with radiations, because it can enhance chiasma farmation, crossing over and chromosomal structural rearrangements; subsequently, as a result of these, easily recognisable traits can be incorporated in an otherwise well adapted variety identified for improvement (Mehetre *et al.*, 1996). Mutation is a vital event influencing karyological variation in evolutionary process, which is reflected on genotypic expression (Gustafsson 1965). Success in mutation breeding depends on mutagenic responsiveness of the genotype concerned (Hageberg *et al.*, 1963).

Induced mutation in *Nigella sativa* L. was totally inexplored till late seventies. Earlier investigations were carried out mostly at chromosomal level involving structural anomalies.

In *N. sativa* Mandal and Basu (1978) studied post-irradiation chromosomal aberrations in leaf meristerms, pollen mother cells and endosperm. The aberrations increased with increase in the treated doses but decreased with lapse of 2-24 hours time after irradiation. Endosperm was suggested to be the most resistant tissue although it had the largest interphase chromosome volume. Kumar and Nizam (1978) assessed both mitotic and meiotic anomalies following dry and presoaked seed treatment with x-ray in *N. sativa*. Pre-soaked seed treatment revealed higher frequency of aberrations than dry seed irradiation. In most of the cases the aberration occurred through disturbances in spindle organisation; in addition, occurrence of dicentric bridges, acentric fragments, ring configuration and micronuclei were encountered. Subsequently in the advanced generations, Kumar and Nizam (1983) isolated certain viable mutants showing abnormal branching pattern and reduced fertility in *N. sativa*. Occurrence of tri-locular and multi-locular fruits with colouration and ornamentation on fruit wall was also recorded by the same authors.

Mutation breeding research in *N. sativa* L, the black cumin, was carried out in the present author's laboratory for about two decades. Investigations were initated in the late seventies when it was evident from review of literatute that black cumin and allied species were neglected by geneticists and plant breeders notwithstanding prospective economic, significance. Possibility of its improvement through mutation breeding was also rather inexplored, and cytogenetic observations were carried out only at primary level although it is a good material for cytogenetic studies, particularly for low number, good stainability and relatively big size of its chromosomes. Primarily endeavour was made to analyse cytogenetic consequences arisen through treatment with different mutagenic agents; subsequent studies involved identification, characterisation, inheritance, protein estimation and combining ability analysis of the isolated mutant lines.

Materials and Methods

Most of the observations were made in *Nigella sativa* L. Seeds were treated with 0.5 per cent , 0.75 per cent and 1.0 per cent ethyl methane sulfonate (EMS) for 2 and 4 hours, 4 Kr to 30 Kr x-ray and 5 Kr to 60 Kr gamma ray irradiations.

Cytogenetic consequences were enumerated by the study of mitotic and meiotic anomalies, morphological modifications, growth and developmental aspects including sterility in pollen, seed and fruit. Mutations were screened cytogenetically and on the basis of chlorophyll deficient variation and other variations due to morphological parameters during M_2 generation. Chlorophgyll mutations were identified following Blixit (1961). Efficiency and effectiveness of the mutagens were determined using the formula used by Konzak *et al.* (1965).

Chi-square test was used for determining mode of inheritance. The true breeding mutants were characterised by quantitative and qualitative evaluation of seed protein profiles, protease and amylase activity and amylase isozyme pattern. Estimation of protein was done following the method of Lowry *et al.* (1951). The techniques of Davis (1964) and Ornstein (1964) were adopted for polyacrylamide gel electrophoresis. The methods of Pai and Gaur (1981), Cardemil (1986), and Salas and Samaddar and Scheffer (1970) were adopted for estimation of protease activity, extraction of amylase isozyme and assay of amylase activity respectively.

Quantitative variations were assessed following usual biometrical techniques. Combining ability was tested adopting the model II method of Griffing (1956).

Results and Discussion

Cytogenetic Consequences

EMS treatment resulted in almost identical chromosomal anomalies in *N. sativa* and *N. damascena* (Datta and Biswas 1981). In both the species mitotic index reduced in 4 hours treatments with all the employed concentrations (0.5 per cent , 0.75 per cent and 1.0 per cent). Chromosomal anomalies increased with decrease in mitotic index. The authors identified in most of the cases a breakage localised in the centromeric region. They also noted that *N. damascena* responded more or less uniformly to all the treatments producing total aberrations, while *N. sativa* was more sensitive to higher concentrations and durations.

Datta and Biswas (1983) estimated sensitivity of *N. sativa* to x-irradiation by analysing post-irradiated (4 Kr–30 Kr) consequences at mitotic and meiotic levels. None of the employed doses was LD_{50}, it was found to be in between 8 Kr and 10 Kr. Chromo–somal anomalies were of different types including fragments, ring configuration, diplochromosomes, laggards with or without bridges, aneuploid variations, unequal length of chromosomes, univalents, multivalents non-orientation, non-synchronous condensation, multipolarity and stickiness. Localised breakages were located at subterminal regions. Total mitotic aberrations increased but mitotic index reduced with increase in the doses used. Micronuclei appeared at interphase. Mitotic disturbances were reported to have affected the physiological process like growth. Pollen sterility and total meiotic anomalies showed parallel relation with doses; this indicated former as the consequence of the latter.

Gamma ray induced cytogentic consequences were studied by estimating seed, germination, seedling growth, cytological irregulaties and pollen as well as seed sterility, in *N. sativa* (Datta *et al.*, 1986). The authors treated dry seeds with 5Kr and 60 Kr gamma ray and ascertained LD50 in between 20 Kr and 30 Kr. Treatment beyond 30 Kr resulted in complete failure of seedling emergence in soil, while 5 Kr irradiation stimulated cell division. Germination frequency, seedling growth and mitotic index reduced but mitotic anomalies increased with increase in the employed doses of gamma-rays. Growth inhibition might be the consequence of mitotic depression resulted due to the induced mitotic anomalies. Lagging fragments and micro-nuclei increased in treatment with higher doses, possibly the latter appeared as the fate of the former. Meiotic anomalies have been ascribed for pollen sterility since frequency of both increased showing parallel dose dependent relation.

Observations of Datta and Biswas (1984a) revealed occurrence of EMS induced fragments, univalents, multivalents (tri, quadri and hexavalents) ring configuration, grouping association, prococious movements, laggards, bridges with or without acentric fragments, cell doubling and asynchronous division in *N. sativa*. Meiotic anomalies have been attributed for pollen sterility in this material also since both enhanced with enhancement in the concentrations of EMS.

Reciprocal translocation was encountered in the meiocytes of M_1 plants raised through gamma ray irradiations in *N. sativas* by Datta and Biswas (1984b). Quadrivalent association was scored in 49.38 per cent PMCs, of which ring configuration was seen in 41.08 per cent cases and the rests appeared in chain. At anaphase I alternate disjunction occurred in 65.55 per cent PMCs and in 34.45 per cent PMCs adjacent separation occurred. The plant was highly pollen sterile (55.8 per cent) showing considerable reduction in seed fertility as compared to normal control plants.

Induced mutagenesis has been studied recently in *N. sativa* by dry and presoaked seed treatments with EMS, gamma ray and H_2O_2 separately and also by combined treatments with two of these mutagens (Rang and Datta 1998, Rang 2000). Almost similar chromosomal anomalies were noted as reported earlier (Datta and Biswas, 1983, 84b, Datta *et al.*, 1986). In combined treatmets with gamma ray + EMS or gamma ray + H_2O_2 Rang (2000) observed higher frequency as well as spectrum of aberrations. In most cases total anomalies increased with doses of the mutagens, while mitotic index reduced. Stimulating effect observed by Rang (l.c). in 5 Kr irradiation in dehydrated seeds coroborated the earlier observation of Datta *et al.* (1986), while in case of presoaked seed treatment extreme mitotic depression was recorded. Pollen sterility increased in all the employed treatments and it was remarkably high in EMS (single and combined) treatments.

Identification and Cytogenetic Characterisation of the Induced Mutations

Datta and Biswas (1985a) studied induced mutagenesis by estimating germinability, survivality, seedling growth, seed yield/ plant, cytological abnormalities, pollen and seed sterility, chlorophyll mutations and other viable mutations in *Nigella sativa* treated with x-ray irradiation and EMS. Altogether 25 types of different mutations including 12 kinds of chlorophyll mutations were isolated by the authors.

Chlorophyll Mutations

The chlorophyll mutations were classified following Blixit (1961). Frequency of the 12 type of chlorophyll mutations followed the order *albina* > *xantha* > *chlorotica* > *chlorina* > *chloroxantha* > *albescens* > *albino–terminalis* > *xantha–terminalis* > *lutea* > *viridis* = *marginata* > *coeruleovirens*. Among these, *marginata*, *chloroxantha*, *coeruleovirens* and *viridis* were viable. *Viridis* and *chloroxantha* were found to be true breeding. Due to reduced germinability, poor survivality, high pollen sterility and meager seed setting it was difficult to raise further generations of the viable chlorophyll mutants and only *viridis* could be studied in subsequent generations.

Mutation for Other Traits

The mutations of this categories were isolated by their specific distinguishing characteristics and classified into 13 different types of mutations including *lax branching, feathery leaf, bushy, male sterile crumpled leaf, dwarf, early flowering, prostrate, brown seed coat, cup leaf, needle leaf, crinkle leaf* and *cotyledonary leaf* mutations (Datta and Biswas 1984c, 1985a). Four of these mutants namely *cup leaf, needle leaf, crinkle leaf* and *cotyledonary leaf* mutants were nonviable.

Lax Branching Mutant

This mutant was identified in the M_2 progeny raised through 2 hours treatment with 0.5 per cent EMS. (Datta and Biswas 1984d, 1985a). It was provided with higher number of branches (48.0) than the control (13.74 ± 1.70). meiotic studies in the mutant revealed occurrence of cytomixis at prophase I; the evidence of which was conspicuous by the presence of intercellular cytoplasmic connections and chromosomal transfer of cytoplasmic contents and chromosomal materials occurred always from a donor pmc to a recipient meiocyte but never in reverse direction. In subsequent stages aneuploid and polyploid variations in chromosome number was recorded in 19.4 per cent cases. Pollen sterility increased slighly in the mutant but seed fertility not affected. In the selfed M_3 progeny of the mutant a trisomic plant (2n+1=13) was detected (Datta and Biswas 1984d). The plant was weak with slender stem and drooping lamina; it failed to produce any seed. Origin of this plant was ascribed to fusion of a normal gamete (n=6) with an abnormal gamete formed as an (n+1=7) outcome of cytomixis.

Another phenotypically aberrant plant was detected in the selfed M_3 progeny of the *lax branching* mutant (Datta and Biswas 1859b). It revealed aneuploid and polyploid variations during meiosis and accompanied with remarkably high pollen sterility and extreme reduction in pollen size. Such meiotic variations occurring in the pmcs of the same microsporophyll have been referred to as 'chromosomal mosaic' and the authors regarded them as the possible consequence of cytomixis.

Feathery Leaf Mutant

Datta and Biswas (1985a, 1986a) isolated this mutant in the M_2 generation raised through 2 hours treatment with 0.5 per cent EMS.

It was detected at the very seedling stage by the characteristic light green colour and prominent feather like pinnae of the leaves (pinnately dissected). Broader and elongated pinnae manifested the feathery appearance. The mutant was taller (49.0 cm) than the control (47.72 ± 1.30). Pollen fertility was quite high (93.72 per cent) albeit occurrence of meiotic anomalies, seed setting was also almost like normal (63.17 ± 4.85 and 65.60 ± 4.24/capsule). It bred true in subsequent generations. In the selfed M_3 progeny of this mutant 4 remarkably dwarf (13 cm-21.6cm) plants with extremely condensed internodes were encountered. The dwarf plants had identical feathery leaves and these were clustered around the stem forming crown like appearance. These plants were designated by the same authors as the *'teles copic mutant'*. Two of the *telescopic* mutants survived and manifested semisterility. Meiotic studies revealed occurrence of ring and chain like quadrivalents, but with high percentage (69.5 per cent) of pollen fertility in the M_3 generation, while in M_2 generation paired fragments were met with. Only 4 plants matured in the M_4 generation manifesting the mutant phenotype. One of these plants revealed ring like quadrivalent in 37.0 per cent meiocytes. The orientation of the quadrivalents in the mutant plant was disjunctional (alternate) in about 43 per cent cases and non-disjunctional in 55 per cent pmes. Mostly anaphasic separation was non-disjunctional and unequal bridges with accompanying lagging fragments were also come across. Phenotypically it was extremely dwarf (11.0cm) with very low seed setting (19.0 seeds/capsule) and considerably high pollen sterility (45.4 per cent). In the other three *telescopic* mutants with dwarf height and feathery leaf phenotype pollen sterility reduced (13.6-15.1 per cent) considerably and seed setting improved (32–65 seeds/capsule) appreciably. The M_4 *telescopic* mutant was semisterile. Origin of this *telescopic* mutant has been attributed to deficiency of genes as an outcome of chromosomal deletion in the parent (Datta and Biswas 1986a).

Bushy Mutant

Dutta and Biswas (1985a,c) detected a bushy plant in the M_2 progeny raised through 2 hours treatment with 0.5 per cent EMS. Higher number (38) of branches and shorter height (35.6 cm) than control provided it with the bushy appearance. The mutant showed conspicuously delayed germination, flowering and maturity with high frequency of pollen sterility and poor seed setting. Meiotically it was irregular with desynaptic behaviour of chromosomes, reduced frequency of chiasma, occurrence of univalents, unequal anaphasic separation accompanied with laggards, and unequal sized microspores in tri-spores and polyspore. The mutant bred true. The characteristic delayed germination, flowering as well as maturity and desynaptic behaviour of chromosomes in the mutant might be due to pleiotropic or linked association of the traits with the bushy phenotype (Datta ad Biswas 1985c).

Early Flowering Mutant

The mutant was detected in the M_2 progeny of 6Kr gamma ray irradiated *N. sativa* (Datta and Biswas 1985a). It flowered and matured earlier than control by 10 days. The mutant plant was taller (50.5 cm) in height without showing any type of lodging tendency. Meiotically it was normal with very high degree of pollen fertility (96.89 per cent). Number of filled seeds per capsule, however, reduced (57.63) as compared to control (65.1). The mutant bred true in subsequent generations.

Dwarf Mutant

This mutant was recognised in the M_2 generation of gamma-ray (8Kr) irradiated *N. sativa* by its remarkable short height (16.5 cm), dark green leaves and infolding of the pinnae (Datta and Biswas 1985a). The mutant flowered after 82 days of sowing, while in the control plants flowering occurred after 70-98 days. Meiosis was normal with only 7.82 per cent pollen sterility. It bred true.

Prostrate Mutant

The mutant was identified by its characteristic prostrate habit in the M_2 progeny raised through 2 hours treatment with 0.75 per cent EMS. Besides prostrate habit leaves were dark green in colour along with delayed flowering as well as maturity and extremely poor see setting (31.27 ± 7.4 seeds/capsule). Meiosis was highly irregular with 79.66 per cent pollen sterility. It bred true (Dutta and Biswas 1985a).

Brown Seed Coat Mutant

It was distinguished from the control and other mutants in the M_2 progeny raised through 2 hours treatment with 0.75 per cent EMS by its characteristic brown colour of seed coat (Datta and Biswas 1985a). The mutant was shorter in height (36.0 cm) than the control and rather poor in yield (47.2/capsule). Meiosis was quite normal in this plant. It was true breeding.

Male Sterile Mutant

This mutant was detected in the M_2 generation following treatment with 0.5 per cent EMS for 4 hours. At maturity the mutant was distinguishable by its characteristic foliage with dark green thick and feathery leaves. Flowering occurred synchronously after 75 days of sowing in place of 70-98 days in control. Meiosis was normal but pollen grains were totally sterile. The probable cause of which is not known. Fruit setting in this plant was possible only through artificial cross pollination with normal control, which indicated that the mutant plant was completely male sterile (Datta and Biswas 1984b, 85a).

The selfed F_2 progeny of the intercrossed F_1 (hybrid) segregated in the ratio 3: 1, when the F_1 was crossed with a heterozygous male parent showing normal phenotype, male sterility appeared in 1: 1 ratio. These clearly indicated monogenic segregation and the authors suggested involvement of a pair of recessive alleles 'ms ms' for male sterility in the mutant.

Crumpled Leaf Mutant

Following irradiation with 4 Kr gamma–ray Datta and Biswas (1985a) isolated two extremely dwarf plants having irregularly lobed reduced pinnae with short petioles in the M_2 generation. The plants were meiotically unstable and highly pollen sterile (84.56 per cent).

Crinkle Leaf Mutant

Several *crinkle leaf* plants were detected in the M_2 progeny of x-irradiated and EMS treated plants of *N. sativa*. Only one of them (isolated through 6 Kr irradiation) could mature. The mutant was extremely dwarf (13.5 cm) with short sized, thick, crinkle and chlorophyll deficient leaves (Datta and Biswas 1984c, 85a). It was late in flowering by 21 days. Meiosis was abnormal showing quadrivalent and univalent associations and laggards. Both chain and ring configurations of the quadrivalents were encountered mostly showing alternate orientation and normal disjunction, but pollen sterility was very high and viable seed formation could not be traced. Defective nature of female gametophytic phytic tissue might be responsible for this (Datta and Biswas 1984c).

Cup Mutant

This type of mutant was detected in M_2 generation at the seedling stage and the mutant plants died well before flowering. The seedlings were very weak and dwarf (9-13 cm) with inward folding of the pinnae conforming cup like appearance.

Needle Leaf Mutant

Detected in M_2 generation the mutant type died at seedling stage. The plants were slender with pinnae modified into finely dissected projection like structure, which provided the leaves with needle like appearance.

Cotyledonary Leaf Mutant

The mutant was distinguishable from other by a broad lanceolate cotyledonary pair of leaves and retarded growth (Datta and Biswas 1985a). In the junction of the cotyledonary leaves finely dissected linear elongated pinnae interwined together to form a globular structure. It died within 15-27 days of emergence.

Mutagenic Effectiveness and Efficiency

Datta and Biswas (1985a) assessed effectiveness and efficiency of the employed doses of the mutagens following Konzak *et al.* (1965). Treatment for 2 hours with 0.5 per cent EMS exhibited maximum mutagenic effectiveness and efficiency among all the employed doses of the two mutagens, and 4 Kr irradiation resulted in highest such effect of the different radiation doses used. Likewise Rang (2000) noted maximum mutagenic effects following 3 hours treatment with 0.25 per cent EMS and 5 Kr irradiation in *N. sativa*.

Genetic Studies

Genetic investigations were performed in the author's laboratory by the study of inheritance pattern of the mutant traits (Datta and Biswas 1985a), electrophoretic evaluation of seed protein, protease and isozyme profiles (Datta *et al.*, 1987, Das *et al.*, 1991), quantitative variation in yield related traits (Datta and Biswas 1986b), combining ability using the technique of diallel analysis, hybrid vigour, cytoplasmic influence, heritability, association of yield parameters and path correlation analysis (Maity 1998).

Mode of Inheritance

Inheritance pattern was studied by analysing selfed progeny and the F_2 progeny raised through crosses between the mutant and normal parents. The mutants *lax branching, feathery leaf, bushy, dwarf, viridis* and *chloroxantha* were true breeding. The F_2 raised through intercrosses with normal parents segregated in 1 : 3 monogenic ratio in all but the two chlorophyll mutations, *Viridis* and *chloroxantha* (Table 3.1). Both these mutants have originated possibly through recessive mutation of two pairs of genes. In the cross *viridis* × normal, F_1 was normal and F_2 progeny segregated into 1 : 15 indicating duplicate factor interaction. The F_1 of *chloroxantha* × normal was normal and F_2 segregation (3 : 13) suggested inhibitory type of interaction.

Table 3.1: F_2 Segregation of Some Mutant Traits Following Crosses Between Normal and Mutant Plants of *N. sativa* L.

Phenotype	*Parents*	F_1	*Number of F_2 Plants Observed*			X_2	*Probability*
Lax branching	Normal × Mutant	Normal	58	16	74	0.451 (3 : 1)	0.60 – 0.50
	Mutant × Normal	Normal	37	9	46	0.724 (3 : 1)	0.50– 0.20
Feathery leaf	Normal × Mutant	Normal	41	12	53	0.157 (3 : 1)	0.80 0.54
	Mutant × Normal	Normal	66	20	86	0.140 (3 : 1)	0.80– 0.55
Bushy	Normal × Mutant	Normal	27	7	34	0.343 (3 : 1)	0.80– 0.54
	Mutant × Normal	Normal	31	6	37	1.523 (3 : 1)	0.50– 0.20
Dwarf	Normal × Mutant	Normal	24	7	31	0.097 (3 : 1)	0.80– 0.50
	Mutant × Normal	Normal	18	7	25	0.12 (3 : 1)	0.80– 0.50
Viridis	Mutant × Normal	Normal	154	14	168	1.013 (15 : 1)	0.50– 0.20
	Normal × Mutant	Normal	50	5	55	0.754 (15 : 1)	0.50– 0.20
Chloroxantha	Mutant × Normal	Normal	42	6	48	1.214 (13 : 3)	0.50– 0.20

Rang (2000) reported 20 types of different mutations including 7 different types of chlorophyll mutation in *N. sativa*. Frequency of the chlorophyll mutations followed the order *chloroxantha* > *viridis* > *albina* > *xantha* > *albescenace* > *lutea* > *chlorotica*. The mutants for other traits included *lax branching*, *feathery leaf*, *lax pinnae*, *bushy* I and II, *early flowering*, *dark reddish brown seed coat*, *yellowish brown seed coat*, *bicolour seed coat*, *needle leaf*, *crumpled leaf*, *cup like* and *male sterile* mutants. The last two types were non-viable.

Rang (2000) analysed inheritance pattern in the different types of mutants isolated by him in *N. sativa* and it was revealed that the mutant traits *bushy*, *feathery leaf*, *lax branching*, *dark reddish brown seed coat* and *yellowish brown seed coat* were governed by a pair of recessive gene; while two pairs of recessive genes were involved in the regulation of *chloroxantha*, *viridis*, *bicolour seed coat*, *cup leaf* and *crumpled leaf* mutations. This observation coroborates the monogenic inheritance of *bushy*, *feathery leaf* and *lax branching* mutant phenotypes and digenic control of *chloroxantha* and *viridis* types of chlorophyll mutations reported earlier by Datta and Biswas (1985a).

Quantitative Variations in Yield Parameters

Yield performance was studied in eight true breeding mutant lines namely, *lax branching*, *feathery leaf bushy*, *early flowering*, *prostrate*, *dwarf*, *brown seed coat* and *viridis* by estimating variability in germination, survivality, plant height, number of branches, capsule length, number of chambers/fruit, number of filled seeds/capsule, seed size (length and thickness), 100 seed weight, harvest index and sterility types (flower, fruit, seed and pollen sterility) in M_2 M_3 and M_4 generations (Datta and Biswas 1986b). Year to year variations in control plants were also assessed. Analysis of variance revealed significant variation (F-value) in most of the traits (Table 3.2). Seed and fruit sterility reduced in most of the mutants in M_4 generation (Table 3.3). Number of branches, number of capsules/plant and seed weight increased conspicuously in *lax branching bushy* and *brown seed coat mutants*. In *feathery leaf* mutant increasing trend in the mean values of all but number of primary branches was market and number of filled seeds increased markedly (Table 3.2).

Assessment of polygenic variations in the different yield related traits due to x-irradiation and EMS treatment in *N. sativa* revealed either reduction or enhancement of the mean values in the mutant

Table 3.2: Mean, F-value, C.D,C.V. and Heritability in Different Parameters of Control and Mutant Plants of M_4 Generation

Samples	*Characters (Mean Values)*										
	Plant Height (cm)	*No. of Primary Branches/ plant*	*Total Branches/ Plant*	*No. of Primary Branches/ Plant*	*No. of Septa Cap-sule*	*Capsule Length (cm)*	*Filled Seed/ Capsule*	*Seed Length (mm)*	*Seed Thickness (mm)*	*100-Seed Weight (gm)*	*Harvest Index (%)*
Control	51.85	6.03	16.37	17.37	5.7	1.28	65.1	1.30	0.92	0.1393	17.68
Lax branching	47.30	7.80	38.50	39.50	5.3	1.36	58.6	1.35	1.03	0.1573	15.12
Feathery leaf	52.27	5.47	17.57	18.57	6.3	1.29	75.9	1.42	0.94	0.1443	18.31
Bushy	31.68	6.97	27.00	28.00	5.3	0.93	29.1	1.46	0.94	0.1712	4.45
Early flowering	36.65	3.53	6.43	7.43	5.4	1.27	56.3	1.38	0.98	0.1414	13.36
Brown seed coat	34.14	4.17	27.50	28.50	5.4	1.12	44.0	1.46	1.07	0.1600	15.29
Viridis	24.25	2.57	3.90	4.90	5.0	1.04	45.9	1.42	0.99	0.1585	17.05
F-value	14.61**	8.41**	6.24**	6.24**	17.99**	9.63**	6.00**	1.84	4.19*	2.38	6.87**
C.D. at 5 per cent	8.80	2.01	14.56	1.456	0.30	0.16	19.28	0.13	0.09	0.02	5.56
C.V. per cent	12.43	21.63	45.04	42.69	3.11	7.47	20.23	5.35	4.67	8.59	21.61
Heritability per cent (board sense)		81.94	71.17	63.59	63.59	84.97	74.09	62.52	22.32	50.77	31.50
66.19											

*: Significant at 5 per cent level **: Significant at 1 per cent level.

Table 3.3: Comparative Account of Different Sterility Types in Control and Mutant Lines in Subsequent Generations

Plant Type	Sterility (%) in Plant Types in Different Generations											
	Flower			Capsule			Pollen			Seed		
	M_2	M_3	M_4	M_2	M_3	M_4	M_2	M_3	M_4	M_2	M_3	M_4
Control	21.83	29.12	27.33	3061	28.07	31.64	3.50	2.67	2.33	29.82	25.40	33.31
Lax branching	11.33	12.24	14.54	45.00	46.28	40.40	6.74	2.31	3.38	41.89	36.50	27.85
Feathery leaf	19.20	10.68	16.75	26.15	20.67	11.81	6.28	6.23	6.19	19.84	14.84	14.22
Bushy	61.52	53.42	47.46	76.92	57.61	53.36	51.52	47.12	61.62	67.22	65.89	63.30
Early flowering	17.63	22.28	22.13	28.42	23.79	25.76	3.11	2.76	0.47	25.82	31.83	26.33
Brown seed coat	28.70	3.92	29.59	38.96	34.43	37.59	11.34	9.70	5.48	54.32	49.73	33.26
Viridis	56.62	42.39	41.78	72.12	64.79	38.31	11.34	12.32	11.49	41.66	34.16	33.00
Prostrate	34.88	26.79	25.68	51.60	34.92	30.38	79.66	38.76	22.93	68.21	44.30	26.63
Dwarf	24.22	19.64	14.29	24.76	24.12	19.23	71.82	8.42	8.12	35.64	29.97	33.92

lines in comparison to the control (Datta and Biswas 1993). Mean values of height, number of branches, and capsule number shifted in positive direction in the treatment with lower doses of x-ray. In EMS treatment mean values reduced in all the traits barring number of fruit chambers in which marginal increase in the mean values was noted in all the concentrations.

Quantitative Estimation and Electrophoretic Characterisation of Protein

Proteins and nucleic acids are the most important components in biological systems. They not only maintain their own specificity but also play key role in genetic regulation of others. For characterisation of 7 different kinds of mutant lines in *N. sativa*, therefore, quantitative estimation of acid and buffer soluble protein, and electrophoretic analysis of seed protein, protease and amylase activity and amylase isozyme patterns were made (Datta *et al.*, 1987, Das *et al.*, 1991). Electrophoretic studies by Datta *et al.* (1987) revealed that there was quantitative as well as qualitiative differences in nature and types of banding patterns among the mutant M_6 lines.

Among the buffer soluble protein bands, band no. 1 was identified in *prostrate, lax branching* and *dwarf mutants;* band no 2 was absent in *prostrate, feathery leaf* and *viridis;* band no. 3 was resolved in *brown seed coat, feathery leaf* and *viridis;* band nos. 4,5,7 and 8 were noted in all the mutant line; band no 9 was absent in *bushy* and *feathery leaf* mutants; band no.10 was located in *bushy, lax branching, feathery leaf* and *dwarf;* band no 11 was absent in *bushy, feathery leaf* and *viridis*, while absence of band no. 12 was marked in *lax branching, prostrate* and *brown seed coat* mutants.

Out of the total number of 12 different bands classified, band number 6 was present in all the mutant lines but prostrate and dwarf mutants. Possibly the protein profiles of band number 6 might have played some role in internode elongation. Remarkable heterogeneity in banding pattern noted among protein content compared to control, buffer soluble protein increased in all but *prostrate* mutant, while barring *lax branching* mutant acid soluble protein content increased in all (Table 3.4).

Out of the 26 types of different acid soluble bands resolved in the control and mutant lines of *N. sativa, brown seed coat, prostrate, lax branching* and *dwarf* mutants demonstrated 13 different types of bands. Ten brands were located in *bushy* and *feathery leaf*, while 11 types were noted in *viridis*. Quantitative variation in buffer and acid soluble protein and distinctive seed protein profiles in different mutant lines provided supportive evidences for induced genetic modifications in the mutant lines.

Table 3.4: Quantitative Analysis of Buffer and Acid Soluble Protein Fractions in Eight Plant Types of *Nigella sativa* L.

Plant Types	*Buffer Soluble Protein Content (gm. Protein/gm. Dry seed)*	*Acid Soluble Protein Content (gm. Protein/gm. Dry seed)*
Control	0.0093	0.0078
Brown seed coat	0.0099	0.0105
Bushy	0.0165	0.0105
Prostrate	0.0084	0.0111
Lax branching	0.0132	0.0069
Feathery leaf	0.0132	0.0081
Dwarf	0.0141	0.0105
Viridis	0.0189	0.0084

Das *et al.* (1991) studied protease and amylase activity and amylase isozyme patterns in the 7 different mutant lines along with the control. Observations on protease activity revealed variation in activity peaks; activities at different pH indicated presence of variety of proteases in the mutant and control plant types tested. This may be attributed to genetic dissimilarities, among the plant types. Amylase activity and variation in amylase isozyme patterns analysed in the mutant and control lines revealed that amylase in mutants differed not only from wild type but also from one another. There were six different amylase spp. demonstrated by polyacrylamide gel electrophoresis.

Electrophoretic analysis of amylase isozymes and the study of protease and amylase activities in thé mutant and control plant types, which are products of gene(s), may enable detection of minor chemical changes of gene(s) that could never be discernible phenotypieally. Since activities of protease and amylase as well as amylase isozyme pattern varied among the different mutant lines with respect to those of control, the authors opined that the genetic background was affected differently in different mutant lines.

Rang (2000) reported higher content of buffer soluble seed protein in the mutant types *feathery leaf bushy* I, *chloroxantha, early flowering, dark reddish brown seed coat* and *yellowish brown seed coat* isolated by him than in the control plants of *N. sativa*.

Biometrical Genetic Studies

Genetic potentiality of some of the true breeding mutant lines of *N. sativa* characterised earlier cytogenetically, genetically and by analysing quantitative as well as qualitative characteristics of protein profiles and isozyme patterns electrophoretically (Datta and Biswas, 1985a, 1986b, Datta *et al.*, 1987, Das *et al.*, 1991) has been further evaluated in the present author's laboratory by estimating combining ability adopting the techniques of full diallel mating design (Maity 1998).

The mutant genotypes including *early flowering, feathery leaf, lax branching* and *viridis* were studied in advanced generation (M_{12}) along with a control variety (Sutton's Local cultivar) for estimating general combining ability (gca), specific combining ability (sca), hybrid vigour, maternal influence and association of characters on the basis of the performances of 9 different yield related traits (Maity 1998). The genotypes were significantly different from one another in respect of plant height, number of primary branches, number of secondary branches days to 50 per cent flowering, days to maturity, number of capsules per plant, capsule length, 100 seed weight and total yield per plant.

The mean squares due to gca were significant for all the traits; mean squares due to sca were also significant for all but number of primary and secondary branches/plant. This indicated that barring number of primary branches all the characters are governed by both additive and non additive components, while number of primary branches was influenced by additive genes only. Analysis of gca effects revealed that *lax branching* mutant was superior to others for transmission of number of secondary branches and capsules per plant, while *early flowering* mutant was the best general combiner for transmission of earliness in flowering as well as maturity and taller height. The Suttons local cultivar was the best general combiner for 100 seed weight and total seed yield/plant, which was followed by *lax branching* mutant.

The best cross combination for specific yield parameter was identified by the analysis of sca for the traits concerned. The cross combination *early flowering* mutant × *lax branching* mutant was most effective for transmission of number of capsules/plant and plant height; for transmission of height, maternal influence also played significant role. The trait was, however, negatively correlated with yield. For transmission of earliness, the best cross combinations were Suttons local × *early flowering* mutant and *lax branching* mutant × *early flowering* mutant. As parent genotype *lax branching* mutant was superior to all for number of capsules/plant and pure line selection would be the effective means for improvement in yield. For transmission of 100 seed weight the best cross combination was *lax branching* mutant × *feathery leaf* mutant. Total seed yield/plant was most effectively transmitted through the cross combination *early flowering* mutant × *viridis* mutant, not withstanding both of them were poor general combiners. The phenomena over dominance and epistasis might be responsible for such performance.

REFERENCES

Abdel, S. I. M., Abdel, W. S., El-Aaser, A. A. and El-Merzabani, M. M. 1992. Biochemical and cytotoxic effects of *Nigella sativa* L. Egypt. J. Biochem. 10(2): 348-355.

Al-Hadar, A., Agel, M. and Hasan, Z. 1993. Hypoglycemic effect of the volatile oil of *Nigella sativa* L. seeds. Int. J. Pharm. Nether. 31(2): 96-100.

Anonymous, 1991. A dictionary of Indian Raw Materials and Industrial Products. *In:* The Wealth of India, CSIR Publication, New Delhi. Vol. VII N-P pp 63-65.

Blixit, S. 1961. Quantitative studies of induced mutations in peas V. Chlorophyll mutations. Agri. Hort. Gent. 19: 402–447.

Chakraborty, H. L. and Chakraborty, D. P. 1964. Spices of India. Indian Agriculturist 3: 158-159.

Das, J. L., Datta, A. K. and Biswas, A. K. 1991. pH differences of protease (s?) and amylase activity and amylase isozymes in control and mutant lines of *Nigella sativa* L. Bangladesh J. Bot. 20(2): 117-123.

Datta, A. K. and Biswas, A. K. 1981. EMS induced mitotic aberrations in *Nigella sativa* L. and *N. damascena* L. Cell & Chr. Newsletter 4: 1–2.

Datta, A. K. and Biswas, A. K. 1983. X-ray sensitivity in *Nigella sativa* L. Cytologia 48: 293–303.

Datta, A. K. and Biswas, A. K. 1984a. Cytomorphological studies in EMS treated *Nigella sativa* L. Pers. Cytol. & Genet. 4: 299–304.

Datta, A. K. and Biswas, A. K. 1984b. Radiation induced translocation in *Nigella sativa* L. Chromosome Information Service (Jap). 37: 10–11.

Datta, A. K. and Biswas, A. K. 1984c. Cytomorphological studies in a *crinkle leaf* mutant of *Nigella sativa* L. Ibid 37: 11-12.

Datta, A. K. and Biswas, A. K. 1984d. Cytomixis and a trisomic in *Nigella sativa* L. Cytologia 49: 437–445.

Datta, A. K. and Biswas, A. K. 1984e. Induced *male sterile* mutant in *Nigella sativa* L. Cell & Chr. Res. 7(1): 24–25.

Datta, A. K. and Biswas, A. K. 1985a. Induced mutagenesis in *Nigella sativa* L. Cytologia 50: 545–562.

Datta, A. K. and Biswas, A. K. 1985b. Meiotic instability in the microsporophyll of an aberrant plant isolated from the mutant progeny of *Nigella sativa* L. Ibid 50: 449–454.

Datta, A. K. and Biswas, A. K. 1985c. A EMS induced *bushy* mutant of *Nigella sativa* L. with desynaptic behaviour of chromosomes. Ibid 50: 535-543.

Datta, A. K. and Biswas, A. K. 1986a. Cytogenetic studies in two induced leaf mutant lines of *Nigella sativa* L. Ibid 51: 309–317.

Datta, A. K. and Biswas, A. K. 1986b. Evaluation of quantitative characteristics in some mutant lines of *Nigella sativa* L. Ibid 51: 298–299.

Datta, A. K., Biswas, A. K. and Sen, S. 1986. Gamma radiation sensitivity in *Nigella sativa* L. Ibid 51: 609–615.

Datta, A. K. and Biswas, A. K. 1993. Induced polygenic mutation in *Nigella sativa* L. Bangladesh J. Botany 22: 89–91.

Davis, B. J. 1964. Disc electrophoresis II. Method and application to human serum proteins. Ann. N. Y. Acad Sci. 121: 404–427.

Griffing, B. 1956. Concept of general and specific combining ability in relation to diallel crossing systems. Aust. J. Bot. Sci. 9: 465–93.

Gustafsson Å. 1965. Characteristics and rate of high productive mutants in diploid barley. Rad. Bot. (Suppl). 5: 323–337.

Hageberg, Å. Persson, G. and Wibarg, A. 1963. Induced mutation in the improvement of self pollinated crops. Recent Breeding Research (Upsale), pp 105–124.

Hailat, N., Bataineh, Z., Lafi, S., Raweily, E., M., Al-Katib, M. and Hanash, S. 1995. Effect of *Nigella sativa* L. volatile oil on Jurkat. T-cell leukemia polypeptide. Int. J. Pharmacogon. 33(1): 16–20.

Konzak, C. F., Nilan, A., Wagner, J. and Foster, R. J. 1965. Efficient chemical mutagenesis. *In* The use of induced mutations in plant breeding. Rad. Bot. 5 (Suppl).: 49–70.

Kumar, P. and Nizam, J. 1978. Effect of x-rays on *Nigella sativa* L. Proc. All Ind. Cong. Cytol. Genet. 3 (Abst) pp 7.

Kumar, P. and Nizam, J. 1983. Experimental mutagenesis in *Nigella sativa* L. Proc. 7th Int. Cong. of Rad. Res. (eds. Broerse J. J., Barendsen G. W., Ked H. B. and Vander Kogel A. J). E 6-14 Amsterdam.

Lowry., O. H., Rosebrough, N. J., Farr, A. L. and Randall, R. J. 1951. Protein measurement with the folin phenol reagent. J. Biol. Chem. 193: 265–275.

Mehetre, S. S. Mahajan, C. R. Shinde, R. B. and Ghatge, R. D. 1996. Assessment of gamma ray induced genetic divergence in M_2 generation of soybean. Ind. J. Gent. 56(2): 186 -190.

Maity, S. 1998. Genetic studies on combining ability in *Nigella sativa* L., Ph. D. thesis, Kalyani University, Kalyani, W. Bengal.

Ornstein, L. 1964. Disc electrophoresis I. Background and theory. Ann. N. Y. Acad, Sci. 121: 321–349.

Pai, K. U. and Gour, B. K. 1981. Effect of pH of the assay medium on protease activity in germinating barley seeds. Ind. J. Pl. Physiol. 24: 168–170.

Rang, S. 2000. Induced mutagenesis in *Nigella sativa* L., Ph. D. thesis Kalyani University, Kalyani W. B.,

Rang, S. and Datta, A. K. 1998. Influence of some physical and chemical factors on the gamma radiation sensitivity in *Nigella sativa* L. (black cumin). J. Natl. Bot. Soc. 52: 17–22.

Rang, S. and Datta, A. K. 1998. A male sterile mutant with desynaptic behaviour of chromosomes in *Nigella sativa* L. J. Phytol. Res. 11(2): 91–94.

Salas, E. and Cardemil, L. 1986. The multiple forms of a–amylase enzyme of the *Araucaria* species of South America: *A. araucana* (Mol). Koch and *A. angustifolia* (Bert). O. Kut Z. Pl. Physiol. 81: 1062–1068.

Samaddar, K. R. and Scheffer, R. P. 1970. Effect of *Helminthosporium victoriae* toxin on germination and aleurone secretion by resistant and susceptible seeds. Pl. Physiol. 45: 586–590.

Sigurbjornssen, B. 1977. Mutation in plant breeding programmes. *In* Mannual of Mutation Breeding, 2nd edn. 1AEA, Viena, pp 1–6.

Scossiroli, R. E. 1965. Value of induced mutations for quantitative characters in plant breeding. Rad. Bot. 5(Suppl).: 443–450.

Watt, G. 1972. A dictionary of the economic products. Cosmo. Publ., Delhi, 5: 428.

Willis, J. C. 1967. A dictionary of the flowering plants and ferns. Cambridge Univ. Press, London.

Chapter 4

Improvement of a Value Added Medicinal Herb "*Trigonella foenum graceum* L.": Need and Approach

S.K. Datta & V.L. Goel

National Botanical Research Institute, Rana Pratap Marg, Lucknow – 2260 01 (U.P.), India

INTRODUCTION

From centuries various plant resources have been used extensively as food, vegetable, fodder, spices and medicines for both preventive as well as curative purposes. Cultivation and improvement of such multipurpose plants has gained immense attention in plantation programs in view of the national goal of country's health care program and for meeting the ever increasing demand to various other products as food, vegetables, condiments, fodder etc. to common man. *Trigonella foenum-graecum* L. commonly known as fenugreek or methi (Family Leguminosae, Sub family Papilionaceae) is possibly one of the oldest medicinal plants known to humankind. It is native to southeastern Europe and west Asia and planted extensively in India. It contains a wide array of biologically and/or pharmacologically active steroids, polyphenolics, and volatiles in its leaves and seeds. It has been used successfully as a forage crop due to nutritional value of its leaves, and several industrial uses have been realized for the galactomannans (polysaccharide mucilage) in its seeds.The young green leaves are used as vegetable before pod formation while the entire shoot portion is consumed as green fodder. It is an important, short duration, multipurpose cash crop of India. It is a good source of oil (rich in iron, vitamins A and D, similar in composition to cod liver oil), proteins and vitamins (A, B1, B2, B3, B5, B6, B9, B12, and D). Being a leguminous plant, root nodules enrich the soil with atmospheric nitrogen while the hay is used as animal feed. The seeds are utilized as condiments and possess several medicinal properties.

Extensive studies have been carried out at our institute and other places on standardization of agro-techniques, and chemical and morphological characterization of various medicinal plants including both herbaceous and woody plants (Goel and Behl 2002, Srivastava *et al.*, 1999, Goel and Datta 1992, Ram and Verma 2000, Sheoran *et al.*, 2000, Huang & Liang 2000). However, only few reports are available on selection and improvement of fenugreek in spite of its great commercially valuable as a medicinal crop (Chandra *et al.*, 2000 Dash and Kole 2000). The present study is a part of this program carried out through induced mutation fenugreek at National Botanical Research Institute.

Genetic improvement work without wide variability in the population is not possible. Intra-specific variations are pre-requisite for initiating any plant improvement program. It forms the basis for selection and improvement. Greater is the variability in the population more are the chances of selection and better is the scope of improvement. *Trigonella foenum-graecum* is a self pollinated crop. Natural diversity among genotypes and populations is limited. Therefore there is a need to create variability among plants with in a population through artificial means.

Physical and chemical Mutagens are well known for their potential of inducing variability in qualitative and quantitative traits including morphological characters. As a result of progress in understanding the roll of induced mutations, a large number of economically useful mutant varieties of several seed and vegetatively propagated crop plants, ornamentals and horticultural plants have been commercially released. The present study aimed to investigate extent and nature of variability induced by physical and chemical mutagens in fenugreek followed by intensive selection of variants and identification of desirable mutants with profuse growth and vogor, heavy fruit/seed production, unique morphological traits and phyto-chemical speciality (oil and diosgenin content) for direct use or for breeding programs in order to optimize gains. The phylogenetic significance of different morphological mutants has been discussed.

Approach

Planting Material and Seed Treatment

The material studied during present investigations comprises dry seeds of fenugreek with average 11.5 per cent moisture content. Methyl methane sulphonate (MMS) and Ethyl methane sulphonate (EMS) were selected as the chemical mutagens while gamma rays were used as the physical mutagens. Seeds were divided in to 15 identical lots for treatment with various doses of chemical and physical mutagens along with one untreated seed lot as control set.

Prior to treatment with chemical mutagens, seeds were soaked in water for two hours followed by treatment with various concentrations (W/V) of MMS (0.02, 0.03, 0.04, 0.05 and 0.06 per cent) and EMS (0.3, 0.4, 0.5 and 0.6 per cent) in phosphate buffer of pH 7 for 6 hours at room temperature (22 ± 3° C). Control (untreated) seeds were soaked in buffer solution for the same period. At the end of treatment, seeds were thoroughly washed in running tap water and dried on filter paper before sowing.

Gamma rays were applied as physical mutagens. Seeds were exposed to various doses *viz.* 30, 35, 40, 45 and 50 Krad gamma rays in a gamma chamber 900, a CO^{60} source available at the institute.

Trials and Design

Treated seeds were sown in field along with respective untreated control seeds. There were three replications in each treatment. Plants were raised in complete randomized block design. Number of seeds per replication and treatment was 150. The plant to plant and row to row distance was 20 cm and 26 cm respectively. The material was sown in at the optimum time (October) using uniform cultural practices as recommended for the crop.

In M_1 data were recorded on seed germination, seedling survival and morphological anomalies. At maturity plant height, number of branches per plant, number of pods per plant, number of grains per pod, length and grain yield per plant were studied. All plants were observed very carefully throughout their life cycle for any change in morphological and other parameters from control involving either the whole plant or a part of plant (chimera). At maturity, seeds from treated plants and control plants were harvested separately. Selected M_1 plants were carried over to M_2 generation following statistically planned randomized block design for screening of variants. Plants showing phenotypic changes were marked separately and their seeds were collected individually. The seeds of each selected variant and control plants were raised in M_3 and subsequent generation for confirming their nature (Variant or mutant). Observations were taken on growth and productivity parameters, stomata and pollen characters, behaviour of chromosomes during mitosis and meiosis, variability in crude oil and diosgenin content and changes in seed extract of control and mutant plants.

Chemical Analysis

Variability in chemical traits of the population was studied in respect to oil and diosgenin content. For estimation of oil, seed powder (25 g) was extracted with petroleum ether (60 to 80°C) in a soxhlet extractor on a water bath for 12 hours. Petroleum ether extract was filtered, the solvent evaporated and residue containing oil was weighed.

For isolation of diosgenin powdered seed samples was defatted by hexane in soxhlet extractor. The defatted seeds were hydrolysed with 3.5 N hydrochloric acid for 3 hours on a water bath. Filtered, washed with distilled water and residue dried. The dried resiudue was extracted with hexane in a soxhlet extractor for eight hours. The hexane extract was filtered, the solvent removed and the crude diosgenin was crystalized with acetone. The crystallized material diosgenin was dried and weighed. Gas liquid Chromatohrapgy technique was standardized for quantitative determination of diosgenin using Chemito-3800 GV model with a flame ionization detector. The flow rate of Nitrogen, the carrier gas, was 85 ml/mt.

Under these conditions the retention time was 9 mts. A callibration curve was obtained from G.L.C. of aliquots of a solution containing diosgenin (5 g) in methyal chloride (1ml) Diosgenin content of unknown samples was obtained from the areas of diosgenin peak in G.L.C. A stainless steel column 5 × 1/8" packed with 3 per cent

Observations and Conclusions

Studies carried out in M_1 generation revealed marginal effect of MMS and gamma rays on seed germination while it decreased after EMS treatment whose concentration was ten times higher than that of MMS treatment in the present study. Both chemical and physical mutagens induced seedling injury, lethality, delay in leaf initiation and pollen sterility (Table 4.1). Different type of morphological abnormality such as bifurcation of cotyledonary leaves, stem fusion, unifoliate to hexafoliate leaves and chlorophyll variegation in leaves were observed in treated plants. Most of the changes were chimeric in nature and did not breed true in subsequent generation. Similar effects of mutagens in first generation after seed treatment, have been reported by earlier workers (Guncle and sparrow 1961, Gaul 1977).

In M_2 generation a population of nearly 4500 plants was investigated. A total of 50 variants were screened from treated population showing modification in growth and branching pattern, leaf, pod, seed, stem characters and pod maturity duration. Out of 50 variants 16 were found to bred true in M_3 and subsequent generations (M4 and M5). These were isolated as mutants. Morphological and chemical traits including chromosomal anomalies were studied in different mutants as detailed below:

Mutants Showing Clustering of Pods

Two mutants (# 1 and 2) were isolated from 0.03 per cent MMS treated population showing emergence of one to six axillary pods instead of one (96 per cent) to two (4 per cent) axillary pods in control plants (Table 4.2). The frequency of one, two, three, four, five and six axillary pods in the mutant was 42.5, 28.8, 23.5, 4.2, 0.9 and 0.2 per cent , respectively. Both of these mutants (# 1 & 2) showed emergence of two funnel shaped fused cotyledonary leaves instead of normal two oval shape cotyledons. In addition to this uni–to tri-foliate leaves and shorter plant height of mutant plants were recorded as compared to control plants. There was an increase in number of branches per plant, 100 seed weight and yield per plant than those of the control.

Table 4.1: Morphological Characteristics of Different Mutants Induced as Result of Treatment with Various Mutagens

Mutagen	*Dose/ Concentration*	*Mutant #*	*Characteristics*
EMS	0.3%	1	Tall, Uni to tetra-foliate leaves with Funnel shaped fused cotyledons, large seeds
	0.3%	2	Tall, Uni to hexa-foliate leaves, bold seeds
	0.3%	3	Blackisk Brown Chocolate seeds
	0.4%	4	Dwarf with compact branching & small seeds
	0.5%	5	Dwarf showing early Flowering
MMS	0.03%	6	Mutant showing 1 to 6 axillary pods
	0.03%	7	Dwarf mutant, uni to trifoliate leaves and Pod clustering
	0.03%	8	Green seed coat colour mutant
	0.04%	9	Leaf mutant with stem faciation
	0.04%	10	Tall with uni–to tri-foliate leaves, large seeds
	0.04%	11	Tall with uni–to tri-foliate leaves, large seeds
	0.04%	12	Tall with uni–to tri-foliate leaves, large seeds
	0.04%	13	Leaf mutant with Uni to tri-foliate leaves,
	0.04%	14	Chlorina Leaf Mutant
	0.06%	15	Green Seed Coat Colour mutant
Gamma rays	40 krad	16	Small Seeded mutant
Control	—	—	Two oval cotyledons, trifoliate leaves, roundstem, yellow seeds

Leaf Mutants

Five mutants (# 9,10,11,12,13) were isolated for change in leaflet number from uni-foliate to hexa-foliate leaves in contrast to control having only trifoliate leaves (Table 4.1). All of these mutants were found in 0.04 per cent MMS treated population In one MMS induced

Table 4.2: Growth Characteristics of Control Plants and Different Mutants (Data presented as graphs)

Mutant #	*Treatment*	*Parameters*							
		Height	*Branches per Plant (No)*	*Pods/ Plant (No)*	*Pod Length (cm)*	*Grains/ Pod (No)*	*Grain Weight (g)*	*Seed Yield/ Plant (g)*	*Remark on morphological trait*
1	MMS (0.04%)	106.01±2.83**	18.30±2.09	33.90±7.07***	5.43±0.37***	4.00±0.39***	1.82±0.01***	1.60±0.13***	Tall with uni- to tri-foliate leaves, large seeds
2	MMS (0.04%)	116.50±3.81**	13.60±0.88	90.00±15.21	6.75±0.20*	7.47±0.39 **	1.91±0.01***	5.30±0.64***	Tall with uni- to tri foliate leaves, large seeds
3	MMS (0.04%)	102.10±3.08*	13.00±1.76	64.64±7.38***	8.17±0.22	8.27±0.44***	1.97±0.01***	4.11±0.68***	Leaf mutant with uni- to tri-foliate leaves, large seeds
4	EMS (0.3%)	112.90±3.30**	14.0±1.0	110.80±5.25	7.38±0.32*	5.63±0.45***	1.81±0.01***	10.15±1.58***	Tall, Uni to hexa-foliate leaves with Funnel shaped fused cotyledons, large seeds
5	EMS (0.3%)	115.60±2.30***	16.30± 1.40***	115.90± 24.74**	7.93±0.29***	8.07±0.47**	1.77±0.04***	9.92±1.49***	Tall, Uni to tetra-oliate leaves, bold seeds

Contd...

Table 4.2–Contd...

Mutant #	*Treatment*	*Parameters*							
		Height	*Branches per Plant (No)*	*Pods/ Plant (No)*	*Pod Length (cm)*	*Grains/ Pod (No)*	*Grain Weight (g)*	*Seed Yield/ Plant (g)*	*Remark on morphological trait*
6	MMS (0.04%)	102.22±4.46	13.00±1.33	58.11±4.26***	8.04±0.31	4.44±0.82***	1.77±0.04***	3.92±1.47 ***	Leaf mutant with Uni to tri-foliate leaves, bold seeds
7	MMS (0.04%)	133.20± 0.11***	6.20± 1.07***	29.00± 6.82***	5.13±0.52***	3.08±0.34***	1.74±0.03***	0.98±0.23***	Tall, Leaf mutant with stem faciation
8	MMS (0.04%)	114.88± 4.52***	5.11± 0.41***	72.12±9.93***	9.54±0.40	10.41±0.70**	1.30±0.03**	7.63±1.47***	Tall, Chlorina Leaf Mutant
9	EMS (0.4%)	50.38± 2.011***	21.13± 3.04***	72.38±6.26***	9.05±0.41	12.38±0.68**	0.91±0.03***	18.40±0.54	Dwarf with compact branching and small seeds
10	EMS (0.5%)	38.25± 1.09***	5.40± 0.48***	12.30±1.72***	7.00±0.36*	7.27±0.38***	1.28±0.06*	1.45±0.22***	Dwarf showing early Flowering mutant
11	MMS (0.03%)	55.3±1.01***	23.9±0.80**	124.5±4.45	10.8±1.02	14.4±0.18	1.70±0.02*	25.5±0.90	Mutant showing 1 to 6 axillary pods

Contd...

Table 4.2–Contd...

Mutant #	Treatment	Parameters							
		Height	*Branches per Plant (No)*	*Pods/ Plant (No)*	*Pod Length (cm)*	*Grains/ Pod (No)*	*Grain Weight (g)*	*Seed Yield/ Plant (g)*	*Remark on morphological trait*
12	MMS (0.03%)	52.71±0.88***	22.66±1.20**	122.33±3.46	10.32±0.82	13.96±0.42	1.65±0.03	23.61±0.62	Dwarf mutant, uni to trifoliate leaves & Pod clustering
13	MMS (0.03%)	103.80±2.68**	7.50±0.48***	104.29±10.86	10.50±0.50	12.38±1.02*	1.49±0.01*	18.36±4.23	Green seed coat colour mutant
14	EMS (0.3%)	72.69±3.80**	11.85±1.02	76.61±6.62***	10.37±0.34	14.44±0.62*	0.98±0.05***	9.34±1.08***	Blackisk Brown Chocolate seeds
15	MMS (0.06%)	93.03±5.48	22.70±1.36*	164.77±5.48*	10.15±0.21	15.12±0.36	1.27±0.03*	25.94±1.99	Green Seed Coat Colour mutant
16	Gamma rays (40 Krad)	97.61±2.67	8.50±0.65**	85.61±4.50***	8.33±0.50	17.43±0.62**	1.17±0.02**	14.06±0.71***	Small Seed mutant
Control	NoTreatment	93.50±1.77	13.17±0.77	121.0±6.89	10.09±1.13	15.02±0.39	1.52±0.03	22.39±1.23	Control plants

mutant (# 9) changes in leaf traits were associated with significantly greater ($P<0.001$) flattening of the stem (stem width = 1.57±0.26 cm) as compared to control plants (0.48±0.05 cm). All these leaf mutants were poor in yield but had large & bold seeds.

In addition to this one mutant (# M 14) was identified in 0.04 per cent MMS treated population showing light green chlorina leaves instead of normal green leaves in control. The mutant was significantly taller than the control but it showed relatively small size of flowers, pods and seeds as compared to control plants. Total seed yield per plant was significantly low (7.6 g) in comparison to control plants (22.39 g).

Seed Mutants

Two mutants (# M-8 & M-15) were identified in 0.03 per cent and 0.06 per cent MMS treated population for unique change in seed coat color from normal yellow to green (Table 4.1) while one mutant (# M-3) was isolated from 0.3 per cent EMS treated population for chocolate color seeds. The productivity of green seed coat color mutant in terms of seed yield was higher than that of the control plants whereas the chocolate seed color mutant showed poor yield.

In addition to this one mutant (# 16) isolated from 40 Krad treated population produced small rectangular seeds as compared to control. Differences were insignificant among control and mutant plants in respect to plant height, internode length, leaf size and pod length. However, in the mutant number of branches, pods per plant, 100 grain weight and yield per plant reduced significantly as compared to control plants (Table 4.2).

Dwarf Mutant

One mutant (# 4) was isolated from 0.4 per cent EMS treated population showing dwarfing of plants and 9 days late flowing. Yield associated traits as number of pods, their length, seeds / pod and seed weight were found to be reduced significantly as compared to control plants.

Mutant for Early Flowering Trait

One mutant (# 5) was isolated from 0.5 per cent EMS treated population which flowered 12 days earlier than the control (untreated) plants. In control flowering was observed after 44 days of sowing whereas it took only 32 days in mutant plants. All plants of this mutant line were reduced in vigor and showed small flowers, pods and seeds and yield as compared to control plants.. Such mutant trait are useful in hybridization program as a parent material for inducing early flowering in a high yielding line.

Vinod, *et al.* (2000) identified a giant mutant in fenugreek. They made a comparative study of this mutant with control normal plants.

Cytological Studies

In both control and all mutants 2n=16 chromosomes were observed in root tip cells mitosis. No chromosomal aberrations were detected in control while chromosomal irregularities such as clumping, early separation, bridges and laggards were observed in 2.3 to 4.4 per cent cells of different mutants.

Meiotic studies showed 8 II at diakinesis and metaphase I in both control and mutants. Separation of chromosome at anaphase I was normal in various mutants and control plants. In some mutants as (M1) non orientation of bivalent was recorded. In addition to this single or double bridges at anaphase I & II were observed in a few mutants as M-9. However, anaphase II was normal in control plants. The frequency of cells with chromosomal aberrations ranged from 1.5 to 4.4 per cent.

Pollen Character

Pollen sterility was higher in mutants as compared to control plants. Pollen grain size (both equatorial diameter and polar axis) was significantly ($p<0.001$) reduced in different mutants as compared to control plants.

Phyto-chemical Studies

Mutants were distinct from the control in their chemical constitution. In control, average oil content was 5.72 per cent whereas in different mutants, it varied from 4.55 to 6.68 per cent. Most of the leaf mutants (7) showed relatively higher oil content than the control plants irrespective of the mutagen used. This shows use of leaf abnormalities as marker for such changes. Pandey etal (2000) reported oil as an important source of nematicidal compound.

Analysis of plants revealed marked differences in diosgenin content among control and mutant plants. Diosgenin content ranged from 0.10 per cent to 0.52 per cent in different mutants while it was 0.35 per cent in control plants. Mutants showing changes in seed coat color were poor in respect to diosgenin content whereas leaf mutants had relatively higher diosgenin synthetic potential. Maximum diosgenin was recorded in chlorina mutant (0.52 per cent). In general higher diosgenin content in different mutants was associated with higher oil content. Oncina *et al.* (2000) studied bioproduction of diosgenin in callus culture of Trigenella foenumgraecum.

Mutants studied during present study were distinct from the control in phenotypic characters as well as in chemical constituents. Modification in more than one trait may either be due to mutations induced at different loci or due to pleiotropic effect of a mutated gene.

Variability in diosgenin content of fenugreek seeds collected from different geographical region was reported by Sharma and Kamal (1982). In the present study different mutants induced as result of physical and chemical mutagens showed such variations in oil and diosgenin content. Earlier such differences were reported to be associated with chromosomal constitution of the plants as diploid or tetraploid nature of the plants (Multani *et al.*, 1983).

Acknowledgements

Authors are grateful to the Director of the institute for providing facilities during the course of this study. Inputs provided by various staff members in chemical analysis work are also gratefully acknowledged.

REFERENCES

Chandra K., Sastry E. V. D. and Singh D. 2000. Genetic variation and character association of seed yield and its component characters in fenugreek. Agricultural Science Digest 20 (2): 93-95.

Dash S. R. and Kole P. C. 2000 Association analysis of seed yield and its components in fenugreek (Trigonella foenum-graecum L).. Crop Research Hisar 20 (3): 449-452.

Goel V.L., & and Datta S.K., 1992. Induced morphological mutations in fenugreek. *JIBS*, 71, 65-68.

Goel, V. L. and Behl, H. M. 2002. Intra-specific variations and germplasm conservation in Neem. In: World Neem conference "Neem 2000" organized at Mumbai during November, 27-30,2002 (Published by: Neem Foundation, Mumbai, India, pp 206-211).

Huang W. Z. and Liang X. 2000. Determination of two flavone glycosides in the seeds of Trigonella foenum-graecum L. from various production localities. Journal of Plant Resources and Environment 9 (4): 53-54.

Multani, D.S., Brar, D.S. and Minocha, J.L. 1983. XV International Congress of Genetics, New Delhi, Dec. 12-21 pp 416.

Oncina R, Botia J M, Rio J, A, del and Ortuno A. 2000. Bioproduction of diosgenin in callus cultures of Trigonella foenum-graecum L. Food Chemistry 70 (4): 489-492.

Pandey R, Kalra A, Tandon S, Mehrotra N, Singh H N and Kumar S. 2000. Essential oils as potent sources of nematicidal compounds in Fenugreek (Trigonella foenum.graecum L).. Journal of Phytopathology 148 (7-8): 501-502.

Ram D and Verma J, P, 2000. Effect of level of phosphorus and potash on the performance of seed yield of.Pusa Early Bunching. fenugreek (Trigonella foenumgraecum). Indian Journal of Agricultural Sciences 70 (12): 866-868.

Sharma, G.S. and Kamal, R. 1982. Indian J. Bot. 5 (1): 58-59.

Sheoran R S, Sharma H C, Panuu P K and Niwas R. 2000. Influence of sowing time and phosphorus on phenology, thermal requirement and yield of fenugreek (Trigonella foenum-graecum L). genotypes. Journal of Spices and Aromatic Crops 9 (1): 43-46.

Srivastava, N., Goel, V. L. and Behl, H. M.1999. Influence of planting density on growth and biomass productivity of *Terminalia Arjuna* under sodic soil sites. Biomass and Bioenergy, 1999, 17: 273-278.

Vinod, Sastry E V D and Sharma R. K. 2000. Comparative studies of normal and a giant mutant in fenugreek (Trigonella foenum-graecum L).. Crop Research Hisar 19 (3): 516-518.

Chapter 5

Role of Experimental Mutagenesis for Genetic Improvement of Peas and Soybean

A. Mehandijev

Institute of Genetics, Bulgarian Academy of Sciences, Sofia 1113, Bulgaria

The experimental mutagenesis has enabled to create a great variety of genetic material in regard to important quantitative and qualitative traits for a short time. It appears to be one of the valuable methods in the genetic improvement of important leguminous crops such as peas and soybean. By applying this method it is expected to solve such problems as resistance to biotic and abiotic stress, lodging resistance, increase of the quantity and improvement of the protein quality as well as to produce new forms with increased yielding potential and increased adaptivity.

By means of inducing and using the mutations five varieties of soybean, three varieties of peas and two varieties of beans were created in Bulgarian. Considerable success was also achieved in the mutation breeding of soft wheat (6 varieties), durum wheat (5 varieties), pepper (6 varieties), lentil (2 varieties).

Along with the success achieved in the field of the mutation breeding there are a great number of problems accompanied by applying the mutagenic factors. Our investigations and those of other countries suggest that the effectiveness of the physical and chemical mutagenes decrease due to accumulating different chromosome aberration. It occurs suppression of the reparation system and the reproductive possibilities of the irradiated organisms as well as high sterility.

When using the different mutagenes in one and the same crop comparatively one-type mutation spectra are obtained as well as considerably low percentage of mutations possessing useful traits. When increasing the dose of the mutagenes applied threshold barriers in the mutation frequency are observed.

The synthesized chemical mutagenes from the group of the so called supermutagenes, distinguishing by a specific operation mechanisms, did not lead to a significant progress in the experimental induction of mutations, though some valuable applied results were obtained.

All this shows that the genome of the plant species is very well protected against the mutagenic influences.

Irrespective of the fact that in the past years a lot of experimental data were collected regarding the specific influence of various mutagenic factors as well as the relatively specific reaction of certain genes, the main problem with the experimental mutagenesis, *i.e.*, controlling and directing the mutation process remains unsolved. Therefore, one of the main problems seems to be the necessity of increasing the effectiveness and the results of the existing mutagenes. This necessitated more effective methods for inducing mutations to be worked out by using combined treatment with physical and chemical mutagenes, as well as by modifying the ionizing radiation effect and also by applying the mutagenic influences in the different phases of the mitotic cycle or the ontogenetic development of the crop. Those three trends appear to be very important directions of investigating and have direct relation to the problem of the mutagenic effectiveness.

Enhancing the investigations on the directions mentioned above will create prerequisites and conditions for step-by-step solving one of the most significant tasks of the contemporary mutation genetics and breeding, namely the development of methods for purposeful obtaining particular types of mutations and their utilization in solving concrete breeding problems.

Yet, one of the forthcoming and important tasks in the field of the experimental mutagenesis remains to be that of increasing the effectiveness of the existing mutagenic factors and developing more effective methods for obtaining greater genetic diversity. The gene pool created may serve for producing new forms for direct utilization or as components for hybridization with the purpose to improve genetically the existing varieties in the present-day agricultural production.

Approaches and Possibilities of Increasing the Mutagenic Effect of the Ionizing Radiation

Amodification of the biological and genetic effect and possibilities of its using in the mutation breeding as theoretical basis of our investigations served the facts established as well as the regularities in the field of the present-day radiobiology. Studies in this area suggested that there are many factors affecting the mutagenic effect of the ionizing radiation. The more significant of them are as follows: the genome, the age, the stage of the cellular or mitotic cycle, the chromosome number, humidity, temperature and the oxygen content (Kallo 1998).

It is also well known that the radiosensitivity of the biological objects and especially the reaction of the plant cells towards the physical mutagenes could considerably change under the influence of a great number of chemical compounds with different action mechanisms. These chemical substances which are applied during the after-radiation period and could alter the radial reaction of the cell are called modifiers. In the most of cases they arouse protective or radiosensitizing effect. They could cause temporary inhibition of all biochemical systems–reparation of DNA, biosynthesis of RNA, synthesis of proteins.

Therefore, one of the approaches for increasing the mutagenic effectiveness of the physical factors such as gamma rays and fast neutrons proves to be the utilization of the modifying effect of many chemical compounds in the form of radioprotectors or radio sensitizers.

The experimental results of the influence of the amino acid DL-threonin on the mutagenic effect of both kinds of reaction-gamma rays and fast neutrons are presented in Table 5.1. As a result of the after-radiation treatment of the pea seeds irradiated by gamma rays with this compound, the mutation frequency has increased threefold (from 1.81 per cent to 5.18 per cent). The coefficient of efficiency has been increased twofold as well (CE–from 1.48 to 3.41).

Table 5.1: Influence of the DL-threonin Amino Acid on the Mutagenic Effect of the Gamma Rays and the Fast Neutrons

Treatment	*Mutation, % Frequency*	*Mutation, Types Number*	*CE*	*Msd/l*
1. Gamma rays100Gy+H_2O	1,81 $p<0.001$	15	1.48	0.057
2. Gamma rays100Gy+ DL-Threonin 10^{-3} M	5.18 $p<0.001$	6	3.36	0.200
3. Fast neutrons 1000 rada+H_2O	1.77 $p<0.001$	20	0.82	0.032
4. Fast neutrons 1000rada +DL–Threonin 10^{-3}M	4.09 $p<0.05$	6	1.81	0.111
5. Control	0.20 $p<0.05$	2		

The same result was achieved with the fast neutrons where the mutation frequency increases from 1.77 per cent to 4.05 per cent . However, in this case an abrupt reduction of the mutation spectrum occurred from 15 to 6 mutation types with pure irradiation by gamma rays and from 20 to 6 mutation types with neutron irradiation, respectively.

The indicator of mutagenic effectiveness is also increased threefold with these after-radiation influences (Msd/L) (Table 5.1).

Hence, as a result of after-radiation treating with DL-threonin it has been found out that most of the plants survive, and the mutation frequency rises consequently.

The fact that the mutation spectrum gets narrower could be explain by the fact that the percentage of the mutations increases at the cost of certain mutation types, for which the after-radiation treatment creates better realization conditions. This is of great theoretical significance for the experimental mutagenesis when the aim is to increase the relative frequency of certain types of mutations for the breeding purposes.

In our experiments with the pea the modifying influence of some bases and base analogues on the mutation spectrum of the gamma rays was studied, which was presented by the two main types of mutations–chlorophyllic and morphological. The after-radiation treatment with uracil, hypoxanthine and caffeine caused considerable changes in the amount of the mutation types as well as in the correlation of the two main types–chlorophyllic and morphological (Table 5.2).

As a result of the modification applied the spectrum varies from 5 to 23 mutation types. The proportion between the two main mutation types has been also changed. For example, in the combination gamma rays+caffeine the chlorophyllic mutants relate to the morphological per one progeny on the average as 2.5:1. In the case of after-radiation treatment with hypoxanthine this proportion is 0.1:1 in favour of the morphological mutants, *i.e.*, the phenotypic expression of one mutation type could change at the cost of another one. The same phenomenon occurred under the influence of the uracil as well (0.6:1). In this case the total amount of the mutation types has also increased(23).

Interesting changes in terms of breeding could be caused in the spectrum of the mutations with other modifying influences as well. For instance, by means of the after-radiation treatment with cytochrome C the mutation frequency has increased threefold when garden pea seeds are irradiated by gamma rays (Table 5.3).

Main mutation types: 1-chlorophyllic; 2-stem type; 3-leaf type; 4-inflorescences and flower organs; 5-morphology of pods; 6-physiological; 7–with useful traits.

The relative part of the seven types of mutations has increased with this kind of influence. However, the main point is, that the part of the mutations with useful traits has been significantly increased (from 0.16 up to 0.46 per cent), as well as the part of the physiological and stem mutations, which affects the architectonics of the plants.

Inducing additive mutation variability under the influence of the compounds tested is an interesting phenomenon, which has been observed in our experiments with pea. This was better expressed by modifying the effect of the gamma rays. In the different treatment variants a significant amount of mutations have been observed that do not occur in the spectrum of the pure irradiation with a particular

Table 5.2: Influence of Certain Bases and Base Analogues on the Mutagenic Effectiveness of the Gamma Rays and Fast Neutrons with Pea

Treatment	Mutants per $100M_2$ Plants	Mutants per One Progeny			Ratio Chlor/Morph Mutant	Mut. Types Number	CE	Mst/L
		Chlor.	Morph.	Total				
Gamma rays 10Kr +H_2O	1.81±0.48 p<0.001	0.08	0.06	0.14	1.3:1	15	1.48	0.057
Gamma rays10 Kr +uracil5.10^{-3}M	2.60±0.25 p<0.001	0.10	0.18	0.28	0.6:1	23	1.49	0.119
Gamma rays10Kr+ hypox.5.10^{-3}M	1.75±0.31 p<0.001	0.02	0.13	0.15	0.1:1	17	1.29	0.060
Gamma rays10Kr+ caffeine10^{-3} M	5.30±1.96 p<0.001	0.10	0.04	0.14	2.5:1	5	3.41	0.325

Table 5.3: Frequency of the Main Types of Mutations when Modifying the Effect of the Gamma Rays with the Garden Pea (number of mutants per 100 M_2 plants)

Treatment	Mutations Total %	Main Mutation Types, %							P
		1	2	3	4	5	6	7	
Gamma rays 0 Kr+H_2O	1.62±0.48	0.90	0.32	0.04	0.04	0.08	0.08	0.16	p< 0.001
Gamma rays 10 Kr+ cytochromeC 5.10^{-3}Mp< 0.001	5.22±0.37	1.46	0.93	0.24	1.63	0.25	0.25	0.46	

dose. For example, under the influence of ATP from 8 chlorophyllic mutations 5 do not occur in the spectrum of 10Kr dose when irradiating garden pea. In the case of after-radiation treatment with cytochrome-C from 8 stem mutations 6 expressed themselves additionally. Similar results are noted with the after-radiation treatment with uracil and AET as well (Table 5.4).

Our studies on the possibilities of modifying the biological and genetic effect of the ionizing radiation show that in this way additive genetic variability with the pea could be induced, *i.e.*, mutations could be obtained which do not occur in the spectrum of the independent irradiating by gamma rays. Depending on the after-radiation treatment applied an increase in the mutation frequency is observed which is not connected with expanding the spectrum of the induced mutations. And vice versa, an abrupt reduction of the spectrum (two-,threefold) could occur, without being accompanied by significant changes in the frequency of the mutations. Hence, quantitative and qualitative changes in the spectrum of the mutations occur under the influence of the modifiers. All this suggests that the processes of arising, establishing and expressing the mutations could be influenced on.

From the 7 groups of chemical modifiers tested those compounds connected with the cell energy increase the mutagenic effectiveness of the ionizing radiation to a greater extent. The combinations with the modifiers appear to be more effective when the physical mutagenes participate with doses being medium in regard to the lethality.

Combined Application of Physical and Chemical Mutagenes

In the field of radiobiology and radiation genetics the phenomenon dose/effect has been investigated most intensively with which the frequency and the spectrum of the induced mutations is connected. For instance, with increasing the irradiation dose a linear relationship of the frequency has been observed in respect to the chromosome aberrations. Regarding the visual mutations the mutation frequency rises with increasing the dose of the gamma rays to a certain level, and afterwards some retention in increasing the quantity of the mutations has been found out, *i.e.*, the curve reaches a plateau. The latter is accompanied by decreasing the mutagenic effectiveness and narrowing the spectrum of the mutations induced. This means that with increasing the irradiation dose we encounter thresholds barriers and when exceeding them negative effects are noted which are connected with decreasing the mutagenic effectiveness (CE) and their results (Msd/L) as well.

Consequently, the dose as a factor, by means of which the effectiveness of the mutagenes used could increase, in this case those are the gamma rays, is of limited importance to the experimental mutagenesis. This gives us good reasons to extend our studies on the interaction of two kinds of mutagenes when applied together or combined and which may result in enhancing their effect. This approach is of great actuality like one of the methods for increasing the effectiveness and the results of the existing mutagenes (Mehandjiev 1985).

Our detailed investigations on the problem suggest that the dose and the concentration of both agents exert a considerable influence on the combined effect of physical and chemical mutagenes. From the experimental data in Table 5.5 it is obvious that by combined treating with gamma rays and ethyl-methane-sulfonate and by increasing the EMS concentration to a certain level the mutagenic effect increases. In the case of independent irradiating the mutation frequency is 1.81 per cent and increases to 7.88 per cent with the after-radiation treatment with EMS-0.2 per cent. Above this concentration the mutagenic effectiveness decreases significantly and the coefficient of efficiency falls from 3.65 to 0.44 when the EMS concentration is 0.8 per cent. In this case the chemical mutagene exerts its inhibiting effect at high concentrations. However, the mutation spectrum proves to be the widest with the after-radiation treatment of 0.1 per cent EMS (33 mutation types). Hence, the combined treatment has caused threefold increasing the number of the mutation types induced which is of great significance to the experimental mutagenesis.

Along with the gamma rays the densely ionizing radiation found application in inducing mutations in the 70s-80s of our century. By irradiating with fast neutrons interesting theoretical and applied results have been achieved. The experiments carried out by us showed that the mutagenic effect of the fast neutrons could be combined with a mutagene widely used such as ethylene-imine (Mehandjiev 1972).

Table 5.4: Changing the Spectrum of the Induced Mutations by Modifying the Gamma Rays Effect with the Pea

Treatment	*Total Type of Mutations*	*Including+ New Mutations* 1	2	3	4	5	6	7	New Rate	%
Gamma rays10Kr+H_2O	15	-	-	-	-	-	-	-	-	-
Gamma rays10Kr+ cytochrome C	24	2	6	2	1	2	1	2	16	66.6
Gamma rays 10 Kr + uracil	23	3	5	1	0	1	1	2	13	56.5
Gamma rays 10 Kr + ATP	21	5	2	2	0	2	1	0	12	57.1
Gamma rays 10 Kr + AET	21	2	5	0	0	3	1	2	13	61.9

Investigated plants in M_2–56199 in number

*Including new mutations: 1-chlorophyllic; 2-stem types; 3-leaf types; 4-inflorescences and flower organs; 5-morphology of pods; 6-physiological; 7–with useful traits

Table 5.5: Mutagenic Effectiveness of the Combined Treatment with Gamma Rays and EMS in Pea

Investigated Indices	*Gamma Rays 40Gy+H_2O*	*Gamma Rays 40Gy+EMS 0.1%*	*Gamma Rays 40Gy+EMS 0.2%*	*Gamma Rays 40GyEMS 0.4%*	*Gamma Rays 40Gy +EMS 0.08%*
1. Mutants per 100M_2 plants	1.21 $p<0.001$	5.66 $p<0.001$	7.88 $p<0.001$	6.78 $p<0.001$	3.48 $p<0.001$
2. Amount of mutation types	12	33	27	26	14
3. CE	0.58	2.65	3.65	1.83	0.44

The results presented in Table 5.6 show that by means of the combined treating with fast neutrons and ethyleneimine a wider mutation spectrum has been induced (25 mutation types). With the independent mutation influences they are 12 and 13 types, respectively.

Table 5.6: Influence of the Combined Treatment on the Mutation Spectrum in Pea

Treatment	*Main Mutation Types, Amount*			
	Chloro-phyllic	*Morpho-logical*	*With Useful Traits*	*Total*
1. Ethylene-imine 0.02% + fast neutrons 1000 rada	9	12	4	25
2. Ethylene-imine 0.02%	5	5	2	12
3. Fast neutrons 1000 rada	4	7	2	13

Analyzing the results in Table 5.6 we may conclude that a total mutation effect has been obtained with regard to number of the mutation types induced. As a matter of fact, in the spectrum of the combined treatment the number of the morphological mutations has risen and mainly of those with useful traits such as forms with increased number of pods, with large pods and seeds, early-maturing and with larger wax coating as well. These mutant forms are of great significance to the mutation breeding with the pea. For example, the mutants with enlarged wax coating on the leaves are of great importance for creating varieties resistant to soil drying. Moreover, a specific mutation effect has been achieved with the combined treatment with ethyleneimine and fast neutrons. From the 25 types of mutations 9 have not been observed in the spectra of EI 0.02 per cent and fast neutrons 1000 rada with the treatment of the pea variety Kubrat-3.

Based on the positive results obtained from the development and application of the combined mutagenic influences in pea we applied them to the soybean as well. The results presented in Table 5.7 show that as a result of the combined treatment with gamma rays and EMS mutation frequency has been obtained which is higher than the additive effect (5.72 per cent as compared to the theoretically expected 4.20 per cent). This is accompanied by an increase of the mutation spectrum (20 types of mutations). The coefficient of efficiency has also risen (CE-0.31).

Table 5.7: Mutagenic Effectiveness of the Combined Treatment with the Soybean Variety Beeson

Treatment	*Mutation Frequency, %* *Observed*	*Expected*	*Mut. Types* *Number*	*CE*
1. Gamma rays 100Gy+H_2O	3.23; p<0.001		13	0.23
2. EMS 0.2 per cent	1.28; p<0.01		10	0.12
3. Gamma rays 100Gy+EMS 0.2%	5.72; p<0.01	4.20	20	0.31
4. Control	0.19		3	

It results from this that when optimal combining the dose and the concentration of the physical and chemical mutagene a superadditive increase of the frequency of the mutations induced could be achieved with the combined treatment.

What are the advantages and the effectiveness of the combined applying physical and chemical mutagenes for the mutation breeding? In this way of mutagenic influence we could achieve:

1. Enhancing the mutation rate which leads to additive or superadditive increasing the frequency of the mutations induced;
2. Inducing quantitative and qualitative changes in the spectrum of mutations, thus increasing or decreasing the frequency of the particular mutation types is observed as well as appearance of mutations which do not occur in the spectra of the individual treatments;
3. Inducing additional genetic variability on the basis of which greater genetic diversity is created;
4. Alteration of important quantitative or qualitative traits in the existing varieties.

These theoretical issues and the developed methods for a more effective induction of mutations created prerequisites for achieving valuable results in the mutation breeding with the pea and the soybean in Bulgaria.

The Role of the Experimental Mutagenesis for the Genetic Improvement of the Pea

In terms of genetics the pea is considered as being the best investigated crop with a very well developed genetic map. It was found out that about 40 per cent of the great amount of genes studied are of importance for the breeding. However, the genetic nature of many agronomical, valuable traits is still hidden and not enough investigated. On the other hand, there is a lack of sufficient information about the mutation of genes, which control significant biological and economical traits. Therefore, terms for the mutagenic effectiveness were introduced into the mutation breeding. According to Ehrenberg and Kohzak the effectiveness of the mutagenes is determined by many factors. For its defining different criteria are used, the more important of which are: survival, sterility, lethality, frequency of chromosome aberrations. According to Gaul and Frimmel (1972), the ratio of the quantity of the chlorophyllic mutations to those criteria specifies the term efficiency, whereas the strength of the mutagenes to induce useful mutation alterations determines the term efficacy. Our investigations with pea suggested that by means of successfully applying the mutagenic factors the frequency could rise and the spectrum of the mutations could also extend. This leads to increasing the coefficient of efficiency as well. However, recently more and more the standpoint has imposed that a mutagenic influence might be consider effective when it increases not only the total mutation frequency but also the mutations with useful traits and when it enlarges the spectrum of the useful alterations.

What problems could be solved when genetically improving the pea by means of inducing mutations?

The habitus or the architectonics, as it is accepted, is much essential for the plant resistance to biotic and abiotic stress. The habitus is dependent on the cultivation aims and a determinant or indeterminant type is preferred as well as a branched or an erect one. With some types of garden pea the climbing habitus is preferred. For the purpose of the canning industry it is required to create varieties with erect habitus resistant to lodging.

In this direction of significance are the pea mutants of the affila tvpe that have leaves modified to tendrils, which lends resistance to the plants. Regarding the seed productivity of these mutants it should be mentioned that it is by 75-80 per cent lower than that of the normal type. They could be a very good donor for transferring lodging resistance.

Of interest are also the dwarf mutants. These mutants could occur spontaneously, yet in our investigations they were noticed more often in the spectrum of the physical mutagenes. Valuable dwarf forms were also produced with the lentil by means of the radiation mutagenesis (Mihov and Mehandjiev 1991, Mihov *et al.*, 1997).

It is known that the dwarf habitus is genetically determined by a single recessive factor. It is characterized by low seed productivity. The pea dwarf mutants produced by us distinguish by larger foliage and better drought resistance.

It has been reported about obtained dwarf forms being of low mitotic activity (Bondar 1975). On the base of using the dwarf mutant forms new genotypes have been created which possess increased physiological activity (Adams 1973).

Considering the influence of the abiotic stress factors on the pea productivity, the necessity of developing new forms arises and they have to be resistant to the constantly changing climatic conditions. Therefore, the physiological stress under the influence of humid or very dry period needs to be studied which has extremely set in during the vegetation and in strong cold winds, hails, dry winds, early or late frosts as well. For the winter pea varieties of importance is to create new genotypes enduring low temperature up to -15°C when a snow-cover is absend. In the way of the experimental mutagenesis new mutant forms have been developed for this type of pea which are known as "rosette"type or as having a prostrate habitus. For the needs of spring forage pea and garden pea as well work on creating new genotypes is in progress by using testing and induction of mutations These genotypes should emerge early in spring at temperatures of 1-3°C and should also be resistant to extreme negative temperatures or strong cold winds.

The creation of early-maturing forms being able to give normal yields while avoiding the high temperatures during flowering is very significant for overcoming the negative influence of the abiotic stress.

The genes controlling that trait have a symbol-gene rf and are located in I chromosome.The investigations on pea by Siderova (1973) concerning the genetic nature of the early-maturing mutants are well known. These mutants are supposed to be multigenic recessive and they could be used in the conventional breeding for creating early-maturing genotypes. In this direction the investigations of Gottschalk and Hussein (1975) are well known and they are also used in the hybridization for genetic improving the pea. In our experiments on modification of the genetic effect of gamma rays it has been found out that as a result of the after-radiation treatment with ATP and cystamine the occurrence of early-maturing mutants is the highest (0.30 and 0.20 per cent , respectively) while with the pure irradiation they are observed to be of 0.08 per cent frequency.

With the combined treatment of fast neutrons with ethylene-imine the frequency of these mutants increases up to 0.85 per cent , while with the neutron irradiation of 1000 rada their frequency is 0.33 per cent (Mehandjiev 1985).

When treating with ethyl-methane-sulfonate Blixt (1967), also noted early-maturing mutants in pea being of higher frequency. Our experiments, when treating with EMS, carried out on garden pea variety Kubrat-3 show that the frequency of the mutants with early flowering and maturing (of 7-8 days) varies from 0.02 to 0.06, while in combination with gamma rays this frequency increases to 0.16 per cent (Mehandjiev 1985).

The high-yielding mutants are also of significance to the mutation breeding. It has been established that they could be obtained by treating with physical and chemical mutagenes. In the literature it has been reported on producing pea mutants with higher productivity (Gottschalk 1966). Reportedly, the mutant type fasciation of the stem is located in IV chromosome–78 locus. In our trials with the variety Kubrat-3 this mutation has been observed with the combined treatment with gamma rays+EMS with frequency of 0.08 per cent

The productive mutants are characterized by a larger number of pods and seeds and depending on the treatment their frequency varies from 0.06 to 0.35 per cent . For example, this type was noted to be of higher frequency with the combined treatment with gamma rays + N-nitroso-N-methyl-urethane (0.33 per cent). With these treatments mutants with large pods and seeds were also obtained, their frequency being from 0.06 to 0.36 per cent , as well as mutants with a larger number of pods per petiole with frequency of 0.08 per cent .

Inducing and using mutations of improved quality is of utmost importance for mutation breeding. In the first place, of interest are the high-protein mutants which contribute to solving the protein problem. In his publications Gottschalk,1983, reported that mutants had been received having high protein content from 25 per cent to 34 per cent . Similar results with the garden pea were also achieved by us (Table 5.8).

Table 5.8: Protein content in induced mutant pea lines

Mutant Line	*Kubrat -3 Control*	*776*	*782*	*792*	*809*	*823*
Protein to dry matter	22.7	29.8	31.7	32.1	32.2	32.6
Protein, % to the control	100	131.0	139.1	136.5	141.3	143.1

These mutant lines considerably exceed the initial variety in regard to the protein content. They also distinguish by higher productivity and higher absolute weight and other valuable properties.

Producing forms resistant to biotic stress is of great importance and especially with the garden pea where the use of pesticides is not recommended. It has been proven, that the pea se[illegible] to be a suitable object for investigating the genetics of disease resistance. According to Peter Mattews resistant genes with the Pisum sativum have been described for 10 fungous pathogens such as Ascochita pisi, Aphanomyces eutieches, Eresiphe pisi, Fusarium oxisporum sp. pisi, Perenospora pisi, Uromyces fabe.

Of great importance is also the resistance to the bacterial pathogens–Pseudomonas syringe and to 7 viruses such as BYMV, CYVV, PEMV, PLRV, PSV.

It is very important to study the inheritance of disease resistance in order to obtain resistant forms by means of mutagenesis.

Investigations in this field show that two genetic systems for resistance of the foliage have been discovered. Four dominant genes are responsible for the resistance of the stems and four genes for the resistance of the leaves.

It has also been found out, that the resistance to Aphanomyces eutieches is determined by a dominant gene with a linkage group 31 and the Ascochyta pisi resistance is determined by a single dominant gene with linkage groups 10 and 11. It is known about the resistance to Perenospora pisi that it is dependent on one dominant gene (rpv-1), and that to Psceudomonas syringe is determined by 5 dominant genes. And the resistance to the virus infection, especially to PSV is conditioned by one recessive gene. The pea resistance to Powdery mildew is of great importance, therefore, to obtain resistance forms by means of experimental mutagenesis is a question of present interest. Some authors report that by irradiating with gamma rays such forms could be obtained more often as compared to the ethylmethanesulfonate (Fad 1983). Mutants resistant to the yellow mosaic virus have been induced in beans and they have been applied in the breeding programs (Tulman Neto *et al.*, 1982). Forms resistant to Uromyces pisi have been produced with the pea in Egypt (Micke 1984).

Another type of mutations finding application in the breeding programs proved to be the alterations in the leaf systems (leaf mutants). With the pea they are known as leafless, semi-leafless, fasciata stems (Kaloo 1988). Besides these mutants in our experiments under the influence of physical and chemical mutagenes some others were also observed such as: acacia in which the tendrils have been transformed into small leaflets; reductus with narrow leaflets; crispa–curled bracts and leafs; obuvatos–reverse ovate leaves; bracteolus–with large bracts; undolatifolius–the leaf is not dentate (Mehandjiev 1985). For the breeding purposes of interest are the affila and acacia mutations. On the base of the first type pea varieties resistant to lodging have been created. According to the investigations carried out by Vassileva (1978) the acacia mutation is characterized by higher assimilation due to the larger foliage. The yield given by it is by 20 per cent higher in comparison with the initial variety.

The investigations on the physiology of the mutations induced have contributed to revealing mutants with alternated physiological qualities. Such mutants have been obtained in many crops such as tomatoes, potatoes, pea (Kaloo 1988).

By irradiating with gamma rays mutants in pea have been produced resistant to the day length, which bloom during the short day much more as compared to the long day Knavel (1967)

It has been reported on produced pea mutants having defective photosynthetic reaction. They contained 80 per cent chlorophyll in the leaves, but produced as much as seven-eight times less oxygen. As a cause the blockade of the main photosynthetic system has been indicated (Gostimcki 1973).

To sum up this chapter it may be pointed out that as a result of the more effective methods for inducing mutations applied by us a great genetic diversity in pea has been created. Strain forms with an increased number of pods, mutants of the afila type have been produced, which possess lodging resistance as well as mutants with larger foliage (acacia type) and semi-dwarf mutants with a normal number of pods. They are valuable as donors for lodging resistance. The mutants with a larger number of pods and seeds, i. e. those with increased biological potential of productivity are of particular importance Using the combined treatment with gamma rays and EMS forms with higher protein content (up to 32 per cent) have been obtained. Mutants with different vegetation period have been also produced: early–maturing (of 7-12 days, earlier than the initial one), medium and late maturing.

Mutant lines resistant to pea weevil / *Bruchus pisi*/have been also created.

As a result of the combined treatment with gamma rays and EMS the new garden pea variety Sredez was developed by us, registered by the State Variety Commission in 1997. It is distinguished by higher productivity, increased protein content (29 per cent) and vitamin C. The new variety is tolerant to Powdery mildew and to Ascochyta.

The Role of the Experimental Mutagenesis of the Genetic Improvement of Soybean

The soybean is one of the four world crops on which the problem of the food supply on the planet depends.

Until recently the improving work on soybean has been mainly performed on the basis of the sexual hybridization. The development of the experimental mutagenesis has enabled a more rapid creation of new genotypes distinguished by early maturity, productivity, resistance to biotic and abiotic stress.

By inducing mutations it is possible to create new genetic diversity which appears to be valuable initial material for the breeding purposes. Genetic improvement of registered and adapted varieties could be also attained with respect to important qualitative and quantitative traits.

Which are the more major directions in the mutation breeding with the soybean?

In the first place that is the increase of productivity, which may be realized by improving the plant architecture. This enables the plants to use the light more effectively. For that purpose it is necessary for the leaves to be located vertically towards the stem.

In this direction the breeding now aims at using dwarf or semi-dwarf compact forms possessing higher number of pods. Such forms have been obtained by us which are in the process of investigating their combining ability.

Another problem related to the alternations in the plant architectonics is to obtain a stem being lower and lodging resistant. For producing stable yields it is necessary to apply rational irrigation system. Unfortunately, the varieties being used at present lodge or are inclined to lodging in irrigating conditions. Therefore, the stem thickening and the development of strong root system will contribute to overcoming this phenomenon. The experimental mutagenesis was used to obtain such genotypes. A mutation with thickened stem and mightily developed root system has been produced by combined treating seeds of the Beeson variety with gamma rays and EMS. On this basis the new variety Biser has been created, registered by the State Variety Commission, which distinguishes by higher productivity and possesses a stem resistant to lodging. This variety is very suitable for growing in irrigating conditions (Mehandjiev 1991).

Krausse,1985 also reported on obtaining mutants with higher productivity. Mutant lines with increased productivity have been obtained in Indonesia by irradiating with gamma rays in doses of 100 to 900 Gy (Hendratno and Sumanggono 1991).

The habitus of the plants is in direct relation to the productivity. Mutants with many branches have been obtained by means of the experimental mutagenesis. Determinant type of soybean has been produced from an indeterminate one using this method. Mutants with fasciation stem have also been obtained which facilitates the mechanized harvesting of the crop.

In view of the geographical location of a particular country the creation of early-maturing varieties is of utmost importance for the soybean production. Super-early varieties with vegetation period of 85-90 days are preferred for some northern countries while for others early and medium-early (120-125 days) are preferred. The early-maturing forms obtained could be used directly or as donors in the combined selection. For that purpose combined mutagenic treatment of the American variety Beeson seeds with gamma rays and ethylmethanesulfonate was applied. A mutation in M_2 has been produced that has a vegetation period 30 days shorter than the initial variety. On this basis the early-maturing mutant variety Boryana has been created (Mehandjiev and Gecheva 1983). By means of the radiation mutagenesis the new varieties Zarya and Zvezda have been also developed possessing shorter vegetation period of 10-12 days and higher productivity. These are in the process of multiplication and releasing in Bulgaria. Much earlier-maturing genotypes of soybean with vegetation period of 90-95 days were created by us in the past years. These valuable forms were obtained by irradiating F_1 hybrids with gamma rays where one of the parents is from the group of the superearly lines. Apart from their high early maturity these forms are characterized by good productivity and large seeds as well. Such lines to be pointed out are 2, 5/17, 123/7. These new genotypes show that they bloom and form seed sets prior to the setting in of extremely high temperatures (above 35°C) and give normal yields.

The resistance to abiotic factors is of vital importance for improving the soybean. To overcome the temperature stress the efforts of the researchers now are directed towards obtaining thermoresistant mutant forms which could endure temperatures above 35°C during the flowering period (Nelson *et al.*, 1991). The method of gametophyte selection could be successfully applied for revealing such mutants. For that purpose in early stages of the development of the male gametophyte the method of the luminescent microscopy is applied and the DNA structure is examined.

The creation of resistance to biotic stress with soybean appears to be a very actual problem of the mutation breeding. There is great necessity of creating forms resistant to the agronomically important diseases such as Pseuchmonas glycinea, mosaic, powdery mildew and the new disease diaporte. Investigations in this direction have been conducted by many authors Bowers 1993, Nasim 1990, Nguen Thi Van *et al.*, 1987. Acquiring such resistance could be achieved either by genetic transfer or in the way of the experimental mutagenesis. Obtaining resistant mutants is very promising because the disease resistance is controlled by qualitative genes. In many cases testing is carried out on a mass scale in an early stage, thus thousands of plants could be studied in M_2.

Testing conducted in a appreciable collection of mutant lines being selected with regard to valuable biological and agronomical properties showed that in the lines B22, B115 and B206 a significant number of plants were revealed having reaction type 0 (from 13.2 to 42.4 per cent). The percentage of plants having reaction type 1 (resistant) is very high as well (Table 5.9).

The investigations carried out by us show that the degree of plant resistance to Pseudomonas glycine could change from a susceptible to medium resistant and resistant type, thus on this basis initial material for the selection may be created.

Mutants could be also obtained having type of attack 0 or 1, which should be directly used for producing new varieties tolerant to diseases (Nguen Thi Van *et al.*, 1987).

Creating resistance to the soybean mosaic virus (SMV) seems to be a serious problem as well. Resistant lines have been produced giving by 19 per cent higher yield (Hashimoto 1985). Resistance to Fusarium oxiporum is also a problem concerning many researchers (Leath and Carroll 1981).

Prathuanwong (1985) as well as Jones and Fatt,1985 report on obtained tolerance to the disease Xanthomonas campeatris p.v. glicins, whereas Smutkupt *et al.* report on tolerance to rusts.

For that purpose the analysis and the selection for such mutations should begin still in the M_1.

Table 5.9: Resistance of Soybean Mutant Lines to Pseudomonas Glycine in the Phase of First Fully Expended Leaf

Mutant Lines	*Studied Plants Number*	*Type of attack*									
		0		*1*		*2*		*3*		*4*	
		Number	*%*	*Number*	*%*	*Number*	*%*	*Number*	*%*	*Number*	*%*
Control	25	—	—	4	16.0	13	52.0	8	32.0	—	—
B22	49	15	30.6	19	38.7	15	30.6	—	—	—	—
B 115	33	14	42.4	9	27.7	11	33.3	—	—	—	—
B148	34	4	11.7	22	64.7	8	23.5	—	—	—	—
B 205	5	5	13.2	29	76.3	9	23.7	—	—	—	—

However, the geneticists are aware of the negative correlation between the grain yield and the protein which appears to be a major obstacle when obtaining high-protein forms. Our studies show that by applying more effective influences it may become possible this correlative linkage to be broken and thus to produce forms with higher protein content, without the latter to be accompanied by decreasing the yield (Lidanski *et al.*, 1984).

Experimental induction of biochemical mutations seems to be very promising when the aim is to obtain new forms with high oil content and improved oil quality–alternation of the linoleic acid, iodine number and the trypsin inhibitor. We produced mutants possessing not only higher oil content but also alternated relation between the saturated and unsaturated fatty acids (Tjankova *et al.*, 1982).

In the past years soybean mutants have been revealed in which the protein form SBTY-A_2 controlling the trypsin inhibition is absent. On the basis of these mutants varieties with good protein digestibility could be created which is of importance for the adequate feeding of the animals.

Due to the absence of natural genetic variability regarding the methionine content in soybean the researchers set their hopes on the experimental mutagenesis for obtaining mutants with higher methionine content. For that purpose it is recommended to use isotope technique with C^{14} marked methionine.

The suitability of the mechanized harvesting is another important direction in the mutation breeding. By irradiating with gamma rays and fast neutrons mutant forms with closed habitus and high position of the first pods (15cm above soil) have been created. Mutants have been also produced having seeds resistant to dehiscence and damage which lack spotting as well. Therefore, the mutagenesis may contribute to alternating the habitus of the plants, their form and the seed colour. Using this method mutants with vertically located leaves have been created which contributes to increasing the effectiveness of the photosynthesis. In this direction very good results have been achieved by Nikolov and Mehandjiev 1987, Alexieva 1991.

The genetic improvement of the soybean is impossible without obtaining abundant initial material for the selection. The foremost task of the mutation breeding is to create new genetic variability with respect to important qualitative and quantitative properties using the conventional and the new methods. For that purpose mutant and hybrid forms should be used as well as induced and new varieties, accessions from geographically remote areas.

What are the major methods by means of which the gene pool of the crop could be enriched and conditions for increasing the mutation breeding could be created? The aims mentioned above may be achieved by using the radiation mutagenesis, chemical mutagenesis; combined applying physical and chemical mutagenes in view of increasing the frequency and expanding the spectrum of the mutations; by irradiating F_1 and F_2 hybrids–through this method an increase of the genetic recombination has been achieved and the hereditary variability has been extended; crossing of mutants with well adapted native varieties; in-vitro mutagenesis on the basis of cell and tissue cultures; using anthers and haploid cells for inducing mutations is also very promising with regard to increasing the mutation frequency and expanding the spectrum,

Irradiating gametes and zygotes reveals great possibilities for the experimental mutagenesis. Treating gametes leads to low frequency of the chromosome aberrations in M_1 s well as to higher mutation frequency, whereas irradiating unicellular zygotes contributes to increasing the genetic diversity. In the genetic improvement of soybean the recombination of mutant genes should be used to a larger extent by crossing with other higher-yielding varieties with the aim to obtain genotypes with valuable biological and agronomic properties.

REFERENCES

Adams, M. W. 1973. Plant architecture and physiological efficiency in the field bean in potentials of field beans and other food legumes in Latin America. Ser Semin, 2E Centro International de Agricultura Tropical, Coli Columbia, 266.

Alexieva, A. 1991. Dwarf soybean mutative strain with vertically arrange leaves and increased protein content. Compt. rendus de L'Academie Bulgarie des Sciences, tome 44,N9.

Blixt S., 1967, Studies of induced mutation in peas, XXI, Agri. Hort. Genet., 25, 112.

Bondar, L. M., Tsitlenok, S. L. 1975. A study of mitoticactivity in inerstem of peh mutant. Tsitol. Genet., 9, 53.

Bowers, G. R. *et al.*, 1993. Inheritance of stem canker resistance in soybeen cultivars croquette and doweling. Crop. Sci., vol. 33, N1, p. 67-70.

Fadl, E. A. 1983. Induced mutations for resistance to rust in beans and peas, Mutat. Breed. Newsl., 21,5.

Gaul H., and Frimmel, G. 1972. Efficiency of mutagenesis in induced mutations and plant improvement. IAEA, Vienna, 121.

Gostimcki S. A., 1973, A new pea mutant reaction, Genetica, v. 9, 166.

Gottschalk W., Wolff G., 1983, Induced Mutations in Plant Breeding, Monograf. Theor. Appl. Genetic, 7, Springer Verlag, Berlin.

Gottschalk, W. 1966. The yield capacity of useful mutants, a critical review of a collection of mutants types of Pisum. Mutations in Plant Breeding, IAEA, Vienna, 85.

Gottschalk, W., Hussein, H. A. Pflanzenzuecht, Z. 1975. v. 74, 265.

Hashimoto K. 1985. Breeding for soybean virus resistance in soybean in tropical and subtropical cropping systems. Asian Vegetable Research and Development Center, Taiwan, Taiwan, 151.

Hendratno, K. and Sumanggono, A. M. R. 1991. Improvement in soybean and mungbeam using induced mutations. Plant Mutation Breeding for Crop. Improvement, vol. 1, p. 77.

Jones, S. and Fatt, 1985. Fate of Xanthomonas campestris infiltrated into soybean leaves: ultrastructural study. Phytopathology, 75, 733.

Kalloo, 1988, Vegetable breeding, v. 1.

Knavel D. E., 1967, A day length-sensitive tall mutant pea, J. Hered., 58, 78.

Krausse, G. W. 1985. Soybean mutation breeding in GDR, Mut. Breed. Newsl., 26,5.

Ledth, S. and Carroll, R. B. 1981. Screening soybean for resistance to Fusarium oxisporum. Phytopathology, 71,(Abstr)., 769

Lidanski T., Mehandjiev A., Gechieva P., 1981, Effect of independent and combined mutagenic treatment on certain quantitative characteristics of the soybean, Compt. Rendus de L'Academie Bulgare des Sci., t. 10, p. 1415.

Mehandjiev A. 1985. Problems of mutagenic efficiency and possibilities to increase it inducing mutations. Doctor dissertation, Sofia.

Mehandjiev A. and Gecheva, 1983. New mutant soybean variety Boriana, Mutation breeding Newsletter, IAEA, Vienna, N23, p. 3.

Mehandjiev, A. 1991. Application of experimental mutagenesis in soybean. Plant Mutation Breeding for Crop Improvement, vol. 1, p. 407.

Mehandjiev, A. and Gecheva, P. 1983. Induced mutations by single and combined mutagenetic treatment, Genetics and Breeding, 16, 2, p. 356-364.

Micke A., 1984, Mutation breeding of grain legumes, Plant and Soil, 88, p. 337-357.

Mihov M., Mehandjiev A., 1991, Investigation of mutation variability in lentil, Plant mutation breeding to Crop improvement, IAEA, v. 1, p. 399-406.

Mihov M., Stojnova M., Mehandjiev A., 1997, Radiation mutagenesis in lentil, 6th National symposium Physics–Agriculture, October, 22-24, Sofia.

Nazim, V. M. and El-Hosary, A. 1990. Included promising mutation in soybean resistant to leaf spot fungi. Int. symposium of plant breeding to crop improvement, Vienna.

Nguen Thi Van 1986. Increasing of soya resistance to Pseudomonas glycinea coerper by means experimental mutagenesis. Doct. dissert.

Nguen Thi Van, Mehandjiev, A. and Kuonovsky, Z. 1987. Utilization of mutagenic factors to enhance soybean resistance to Pseudomonas glycinea. Compt. rendus de L'Academie Bulgarie des Sciences, tome 40, p. 113-115.

Nicoclov C. and Mehandjiev, A. 1987, Achievements of mutation breeding in soybean. Experimental mutagenesis in plants, Plovdiv, October 26-30, p. 73-77.

Prathuangwong, S. 1985. Soybean bacterial pustule reach in Thailand, Aciar, Proc. Series N8, 40.

Siderova, K. K. and Uzhintsova, L. P. 1973. Study of morphological characteristics and genetic nature of early mutant of pea. Itogi Nauch Rabot Inst Tsitol Genet, Novosibirsk.

Siderova, K., Khanghild, V.V. and Debely, G. A. 1969. The experimental mutagenesis in pea breeding, Agrobiologia, 4, 538.

Smutkupt S., Wongpiyasitid, A. and Lamseejan, S. 1986. Improvement of rust resistance in soybean by mutation breeding. Improvement of grain legume production using induced mutations, IAEA, p. 371-381.

Tjankova L., Mehandjiev A., Stoyanova Z., Gecheva P., 1982, Biochemical studies on mutants, Compt. Rendus de L'Academie Bulgare des Sci., t.35, p. 1557.

Tulmann Neto, A., Mentana, J.O.M. and Ando, A. 1982. Induction of mutation in French bean breeding in Anais, Expesa Brasileira de Pesquisa Agropecuaria, Capat, Brazil, 25.

Vassileva, M. 1978. Induced genetic variety in P. sativum, in Experimental Mutagenesis in Plants, Sofia, Bulgaria, 440.

Wyatt, J. E. 1984. An indeluscentanther mutant in the common lean, J. Am. Soc. Hortic. Sci., 109.

Chapter 6

Cowpea Mutation Breeding for Resistance to Bacterial Leaf Blight Disease (*Xanthomonas vignicola* Burk.)

Sanit Luadthong

Department of Agronomy, Faculty of Agriculture, Khon Kaen University, Khon Kaen, 40002, Thailand

ABSTRACT

Red Cowpea 6-1 U.S. seed at 12 per cent moisture contents were divided into 5 sets with 3600 seeds each. They were treated with Gammaray at 20, 40, 60, 80 and 100 kR in 1986. The M1 seeds were planted and selection for bacterial leaf blight resistance and seed yield, after M7 selections 58 mutant lines were selected from the treatments at 20, 40, 60 and 80 kR for preliminary yield trial. In early rainy season(May),1989., 58 mutant lines were planted for preliminary yield trial, they were tested in RCBD and 43 mutant lines were selected for planting in late rainy season(August),1989. In 1991, 25 mutant lines were tested for seed yield and resistant to bacterial leaf blight disease, after harvesting 12 mutant lines were selected and planted in dry season(January)1992. In August,1992., 7 desirable mutants were planted for seed yield and resistant to bacterial leaf blight disease. After harvesting 5 desirable mutants with high yield and resistant to bacterial leaf blight were selected for further yield trials.

Advance yield trial were carried out in 3 locations in, 1993, they were tested in RCBD and found that the mutants KKM 40-4-1-1-1, KKM 60-2-1-1 and KKM 40-5-2-1-1 yielded higher than a check ranging from 7-20 per cent . In comparison to the original parent, all the mutants yielded higher ranging from 17-110 per cent . Their seeds were larger than the original parent ranging from 27-81 per cent and found that the mutants KKM 40-4-1-1-1, KKM 60-2-1-1-1, KKM 40-5-2-1-1 and KKM 80-9-2-2-1 resistant to bacterial leaf blight disease. It was found that higher yield were associated to higher number of pod per plant, 100 seeds weight, shelling percentage and resistant to bacterial leaf blight and other diseases.

INTRODUCTION

This report is a part of the project on "The Improvement of Food and Agriculture by Using the Nuclear Technique". The purpose of this research is to find out bacterial leaf blight disease resistance in Red Cowpea 6-1 U.S. variety. This variety is determinate plant type, early maturity, nonsensitive to daylength, red-shining seedcoat, high percentage of germination but susceptible to bacterial leaf blight disease. Conventional method, by crossing to VITA-3 and IT 82 D-889 were made in 1983 and 1984 respectively, however the F4 and F5 seed colour are not red and shining as the ariginal plant. Therefore, by using radiation breeding will be another methods to improve disease resistance and retaining seed colour in this variety.

Material and Methods

Red Cowpea 6-1 U.S. seed at 12 per cent moisture content were divided into 5 lots with about 3,600 seeds each. They were treated with gamma-rays at Faculty of Science, KasetsartUniversity, at 20, 40, 60, 80 and 100 kR. The treated seed (M_1) were planted in green house and experimental field in August 1986 by planting Red Cowpea as a check. They were studied for days from planting to germination, percentages of germination, rate of germination, chlorophyll abnormal, seedling height, days to flowering and days to maturity. At harvesting 3 pods from mainstem of each plant from each treatment were taken and bulked to use as seeded for planting in M_2 selection.

In January, 1987, seeds of the M_2 from each group were planted in row, 5 meter long by planting the check in every 10 rows. However, materials in M_2 were not segregated. The M_3 materials were planted in August, 1987., and mutant lines were segregated and selection for bacterial leaf blight resistance were made.

Result and Discussion

From the Table 6.1 found that days from planting to germination on M1 seeds were delayed when the amount of radiation doses increase. It was found that untreated seeds germinate in 4 days after planting, while the treated seeds germination delayed ranging from 0.5-6 days.

Table 6.1: Comparison Characteristics of Cowpea Seedling in M_1 Treated with Gamma Ray at Difference Doses

		Treated Seeds at				
	Control	*20kR*	*40kR*	*60kR*	*80kR*	*100kR*
Days to germination	4	4.5	5.5	7	8.5	10
Germination ((%)	100	92	79	65	42	32
Chlorophyll abnormality (%)	0	8	21	35	58	68
Seedling height (cm).	15	10	7	5	2	1.5
Days to first flowering	32	35	37	39.5	42	44
Days to maturity	58	63	65	68	70.5	73

Germination of M_1 seeds were decreased as the rate of radiation doses increase. In comparison to the check germination of M_1 seeds at 20, 40, 60, 80 and 100 kR. were reduced by 8, 21, 35, 58 and 68 per cent respectively.

Seedling height at 10 days after planting were also retarded as compares to check. It was found that higher the radiation rate, decrease the seedling height.

Chlorophyll abnormality found to be associated to radiation doses, when increase rate of radiation doses, plants abnormality of chlorophyll were also increased. All the abnormality seedlings died within 3-7 days after emerging from the soil, while the remaining seedlings were stunted growth and some of the flower buds changed into smaller leaves. It was found that, those drying seedlings could not produce terminal shoot after unifoliolage leaf appears.

Days to first flowering and maturity were also affected by rate of radiation, increases rate of radiation doses resulting delayed flowering and maturity. These delays might be due to slow germination rate of seedling and abnormality of chlorophyll in the remaining plants.

During 1988-1989, pedigree selections were carried out on M_4, M_5, M_6 and M_7 in order to increase the cycle of selections. The materials consists the cowpea lines treated with Gamma ray at 20, 40, 60, and 80 kR. The number of plant selections in each generation were presented in Table 6.2 and selection methods were made as follows.

Table 6.2: Number of mutant Lines Selected from Each Treatment in the M4, M5, M6 and M7 During Planting in 1988-1989

Treatments Doses (kR)	*Number of Mutant Lines Selected in*			
	M_4	*M_5*	*M_6*	*M_7*
20	21	44	39	14
40	11	20	21	11
60	17	32	25	12
80	20	47	48	21
Total	**69**	**143**	**133**	**58**

The M_4 materials were planted in February, 1988 consisted 21 lines of 20 kR, 11 lines from 40 kR, 17 lines from 60 kR and 20 lines from 80 kR. They were planted in a single row plot, 5 meters long with 50 cm between rows and 20 cm between plants by planting the check variety (Red Cowpea 6-1 U.S). in every ten rows. The M_4 plants were segregated into white, light green and green pod colours, they were better than the check in plant high, pod size, seed size and resistant to diseases. At harvesting, 143 plants with desirable characters were selected for planting in the next generations.

The M_5 were planted in June, 1988 comprised 44 lines of 20 kR, 20 lines from 40 kR and 22 lines from 60 kR and 47 lines from 80 kR. They were planted in two row plots, 5 meters long with 50 cm. between plants and 20 cm. between rows by planted Red Cowpea 6-1 U.S. in every ten rows as a check. Similar to the M_4, all the M_5 materials were not stable and segregated into early and late maturity. Some of the mutants at 80 kR, were uniformity in plant high and early maturity. While, some of late maturity mutant lines were resistant to bacterial leaf blight and other diseases. At harvesting 133 lines of desirable plants were selected for planting in the next generations.

The M_6 materials were planted in September, 1988 consisted 39 lines of 20 kR, 21 lines from 40 kR, 25 lines from 60 kR and 48 lines from 80 kR. They were planted in two row plots, 5 meters long with 50 cm. between rows and 20 cm. between plants by planting Red Cowpea 6-1 U.S. as a check variety in every 10 rows. The M_6 mutants were uniformity in plant high and pod size, they showed resistant to root rot and other diseases. Their pods are larger and longer than the check variety. They have larger seed size with kaki, red and dark red colours. At harvesting 90 lines with desirable characters were selected for planting in the next generations.

The M_7 materials were planted in January, 1989 consisted 90 lines. They were planted in 2 row-plots, 5 meters long with 50 cm. between rows and 20 cm between plants by planting Red Cowpea 6-1 U.S. as check variety in every 10 rows. Because of dry weather during planting, all of the plants were attacked by aphid and transmited the virus diseases to the plant. At harvesting 58 mutant lines which freed from virus diseases were selected for preliminary yield testing in May, 1989.

In early rain season(May) and late rainy season(August),1989, two preliminary yield trials were carried out on the mutant lines selected from 20, 40, 60 and 80 kR. Each radiated group of mutant was tested separately from one another and tested in a Randomize complete block design. Each mutant line was planted in 4 rows plot, 5 meters long with 50 cm. between rows and 20 cm. between plants by planting Red Cowpea 6-1 US. and KKU. 402 as a check. The characteristic studies were days to 50 per cent flowering, number of pod/plant, number of seed/pod, 100 seeds weight, shelling percentage and seed yield by measuring from 2 central rows.

The results of each treatment was discussed separately from one another. However, the results of each treatment planted in the early late rainy seasons were combined and discussed together.

The mean seed yield of the mutants treated at 20 kR. are presented in Table 6.3. The grand mean in the early rainy season was 1,113 kg/ha. There were seven mutant lines got higher yield than the average and the KKM 20-2-1-1 got the highest seed yield. There were eight mutants lines yielded higher than the original parent, however only six of them were significantly differences from the original parent. The check (KKU 402) gave the lowest seed yield.

Table 6.3: Grain Yields (kg/ha) and Certain Agronomic Characters of Cowpea Treated at 20 kR Grown During the Early and Late Rainy Seasons in 1989

Mutant Lines	*Yield (kg/ha)*		*DFF*	*100 Seed (gm)*	*Pods/ Plant*	*Shelling (%)*	*Disease (1-5)*
	Early	*Late*					
KKM20-7-2-1-1	1701AB	1932A	38	14.15	19	75.74	1.5
KKM20-2-1-1	1882 A	1741 AB	37	14.88	22	80.50	1.0
KKM20-17-2-1-1	1497 BC	1926 A	36	16.89	25	73.17	2.0
KKM20-21-3-2-1	1394B-D	1761 AB	36	15.47	23	73.77	R
KKM20-3-2-1-1	1166 C-E	1988 A	36	23.65	15	76.49	0.5
KKM2017-1-1-2	1369B-D	1738 AB	37	17.73	18	70.84	0.5
KKM20-17-1-1-1	1481BC	1610A-C	36	15.87	18	71.63	2.0*
KKM20-21-3-1-1	922 E-G	1661A-C	37	20.14	23	70.93	1.5
KKM20-9-1-1-1	931 E-G	1420BC	39	20.49	22	79.81	1.0
KKM20-15-2-1-1	1072 D-F	1263 C	37	15.74	18	73.44	2.5
KKM20-21-2-1-2	650 G	1631A-C	36	26.33	20	75.67	1.0
Red Cowpea 6-1U.S.	976 E-G	1243 C	35	12.44	20	74.48	S
KKU 402 (check)	348 H	1391 BC	39	24.28	19	80.42	R
KKM20-21-3-1-1	862 E-G	—	42	—	—	—	—
KKM20-7-1-3-1	791 FG	—	37	—	—	—	—
KKM-19-1-1-2	722 FG	—	35	—	—	—	—
Means	1113	1639	36	17.10	20.15	20.15	—
CV. (%)	18.85	16.22	—	—	—	—	—

* = Cercospora disease. DFF = Day to 50 per cent flowering.

R = Resistant. S = Susceptable.

The mean seed yield of the late rainy season planting was higher than the early rainy planting by 47.26 per cent and all the mutants yielded higher than the original parent. There were ten mutants yielded higher than the check (KKU 402). The mutant KKM 20-3-2-1-1 got the highest yield, however, this mutant had been given low yield in the early rainy yield trial. The mutant KKM 20-7-2-1-1 and KKM 20-2-1-1 gave the highest mean seed yield in both the seasons indicated that these mutants were superior for yield in both the environments. Higher seed yields were associated to the number of pods/plant, 100 seeds weight and shelling percentages.

The mean seed yields of the mutants treated at 40 kR are presented in Table 6.4. The grand mean of the early rainy season was 1013 kg/ha. There were eight mutant lines got higher seed

Table 6.4: Grain Yields (kg/ha) and Certain Agronomic Characters of Cowpea Treated at 40 kR Grown During the Early and Late Rainy Seasons in 1989

Mutant Lines	*Yield (kg/ha)*		*DFF*	*100 Seed (gm)*	*Pods/ Plant*	*Shelling (%)*	*Disease (1-5)*
	Early	*Late*					
KKM40-3-2-1-1	1734 A	1588A	35	17.29	22	75.82	R
KKM40-3-3-1-1	1499 A	1418AB	35	12.27	23	74.05	R
KKM40-4-1-1-1	1132 BC	1603A	35	15.86	20	75.23	1.5
KKM40-5-2-1-1	1078 B-D	1594 A	36	12.29	16	61.85	R
KKM40-9-1-1-1	1156 B	1456AB	35	15.46	22	70.85	1.0*
KKU 402 (check)	876 B-E	1656A	37	24.25	22	79.60	R
KKM40-5-2-2-1	946 B-D	1445AB	37	13.37	23	65.93	1.5*
KKM40-4-2-2-1	921 B-D	1349AB	37	14.52	21	72.59	2.5*
KKM40-7-1-1-1	794 C-E	1186B	38	11.19	18	65.80	R
KKM40-8-1-2-1	908 B-D	—	41	—	—	—	—
Red Cowpea 6-1U.S.	814 B-E	833 C	35	12.16	16	73.72	S
KKM40-8-1-1-1	765 DE	—	36	—	—	—	—
KKM40-3-2-1-1	550 E	—	36	—	—	—	—
Means	1013	1413	36	14.84	20.30	20.30	71.54
CV. (%)	20.74	16.36	—	—	—	—	—

* = Cercospora disease. DFF = Day to 50 per cent flowering.

R = Resistant. S = Susceptable.

yields than the original parent, however, only two mutant lines were yielded highly significant differences from the original parent and the check (KKU 402).

The grand mean of seed yields in the late rainy season was higher than the early rainy season about 39.49 per cent and found that the check (KKU 402) got the highest seed yield, while the original parent got the lowest seed yield. The combines seed yield of both the seasons found that, the mutant KKU 40-8-2-1-1 got the highest average seed yield followed by the mutant KKM 40-3-3-1-1. Higher seed yield in these two mutants were related to the number of pods/ plant, shelling percentage, 100 seeds weight and resistant to the diseases.

The mean seed yields of the mutants treated at 60 kR are presented in Table 6.5. The grand mean in the early rainy season

Table 6.5: Grain Yields (kg/ha) and Certain Agronomic Characters of Cowpea Treated at 60 kR Grown During the Early and Late Rainy Seasons in 1989

Mutant Lines	*Yield (kg/ha)*		*DFF*	*100 Seed (gm)*	*Pods/ Plant*	*Shelling (%)*	*Disease (1-5)*
	Early	*Late*					
KKM60-6-2-1-1	1770A	1854ABC	35	16.02	25	71.81	2.0
KKM60-10-1-1-1	1331BC	2096AB	40	14.24	26	72.38	R
KKM60-1-1-1-1	1486 AB	1906ABC	35	13.36	23	67.18	1.0
KKM60-6-2-1-2	1491 AB	1776BC	35	13.94	23	37.81	R
KKM60-16-2-1-1	1365 BC	1726 BC	36	14.98	14	67.83	1.0
KKM60-16-1-1-1	1484 AB	1599 CD	36	11.98	34	74.98	R
KKM60-6-1-2-2	1449 AB	1619 CD	35	14.35	20	68.47	1.5
KKU 402 (check)	1208CD	1853ABC	37	24.56	24	78.57	R
KKM60-14-1-2-1	1020CDE	1756BC	38	13.17	23	73.46	1.5
KKM60-9-1-1-1	1128BCD	1565 CD	36	11.33	25	63.76	R
KKM60-6-1-1-1	1274BC	1068 F	35	16.73	25	69.47	0.5
KKM60-16-3-1-1	828 DE	1388 DE	36	17.08	20	70.45	1.5
Red Cowpea 6-1 U.S.	704 E	1186 EF	35	12.23	19	75.19	S
KKM60-13-2-1-2	846 DE	—	41	—	—	—	—
Means	1235	1680	36.6	14.38	23.21	71.06	-
CV. (%)	19.20	12.47	—	—	—	—	—

was 1,235 kg/ha and all the mutants yielded higher than the original parent. The mutant KKM 60-6-2-1-1 got the highest seed yield and highly significant difference from the check. Higher yield in this mutant might be related to larger seed and higher number of pod/ plant.

The mean seed yield in late rainy season trial was higher than the early rainy season about 36 per cent with the grand mean 1,680 kg/ha. Most of the mutants yielded higher than the original parent. There were two mutants (KKM 60-12-2-1-1 and KKM 60-10-1-1-1) yielded higher than 2 tons/ha. However, they were not significant differences from the check. These two mutants got moderately yielded in early rainy season, therefore they might be suitable for planting only in late rainy season. The mutant KKM 60-6-2-1-1 had been given highest yield in the early rainy season was also got higher yield in this season indicated that this mutant might be suitable to plant in both the seasons. However, they will be further testing in the early and the late rainy seasons in 1990. The average seed yield in both the season plantings found that the mutant KKM 60-6-2-1-1 got the highest yield followed by the KKM 60-10-1-1-1 and the KKM 60-1-1-1-1. Their high yields were associated to the number of pods/ plant, 100 seed weight and shelling percentages.

The mean seed yields of mutants treated at 80 kR are presented in Table 6.6 with the early rainy season grand mean was1,256 kg/ ha and found that all the mutants yielded higher than the original parent. The mutant KKM 80-13-2-1-1 got highest seed yield and highly significant differences from the original parent and the check (KKU 402).

In late rainy season yield trial, the grand mean of seed yields was1,593 kg/ha which higher than the early rainy season about 27 per cent and all the mutants yielded higher than the original parent. The mutant KKU 80-9-2-2-1 got the highest seed yield, however, there was not significant difference from the check. The combines seed yields of the two seasons showed the mutants KKM 80-17-1-1 got the highest yield followed by the KKM 80-9-2-2-1. Higher seed yield of these mutants were related to the number of pods/plant, shelling percentages and 100 seeds weight.

The study on days to 50 per cent flowering showed that all of the treatments had average days to 50 per cent flowering between 36

Table 6.6: Grain Yields (kg/ha) and Certain Agronomic Characters of Cowpea Treated at 80 kR Grown During the Early and Late Rainy Seasons in 1989

Mutant Lines	*Yield (kg/ha)*		*DFF*	*100 Seed (gm)*	*Pods/ Plant*	*Shelling (%)*	*Disease (1-5)*
	Early	*Late*					
KKM80-17-1-1-1	1766AB	1876AB	40	15.28	26	81.15	1
KKM80-9-2-2-1	1517 A-E	1958 A	36	13.73	23	71.63	2.5*
KKM80-13-3-1-1	1669A-C	1701A-C	38	11.57	26	71.24	R
KKM80-13-2-1-1	1823A	1473B-D	36	13.37	22	70.26	1.5
KKM80-8-1-1-1	1300B-H	1951 A	35	12.79	24	71.70	1.5
KKM80-4-1-1-1	1438A-G	1697A-C	35	15.28	22	70.12	1.5
KKM80-4-4-1-1	1508 A-E	1566A-D	37	13.12	22	74.73	2.0*
KKM80-6-1-1-1	1183 C-I	1783A-C	35	12.96	28	71.65	R
KKM80-23-1-1-1	1258B-H	1694A-C	37	15.33	23	71.62	R
KKU 402 (check)	1368B-H	1552A-D	38	24.38	23	72.16	R
KKM80-9-2-1-1	1141 D-I	1569A-C	35	13.53	24	66.09	R
KKM80-4-3-1-1	1374B-H	1191D	37	11.32	25	75.14	1.5
KKM80-17-2-1-1	1048D-I	1464B-D	35	15.36	23	69.25	2
KKM80-17-3-1-1	898 H-J	1411CD	35	15.08	21	63.94	1
Red Cowpea 6-1U.S.	499 J	926 E	35	12.26	18	76.31	S
KKM80-11-2-1-1	1528 A-E	—	37	—	—	—	—
KKM80-9-1-1-1	1448 A-F	—	35	—	—	—	—
KKM80-12-1-2-1	1384B-H	—	36	—	—	—	—
KKM80-23-1-3-1	1114D-I	—	36	—	—	—	—
KKM80-4-2-1-1	1006 E-I	—	35	—	—	—	—
KKM80-12-3-2-1	951 F-J	—	40	—	—	—	—
KKM80-22-2-2-1	929 G-J	—	37	—	—	—	—
KKM80-12-1-2-1	743 I-J	—	41	—	—	—	—
Means	1256	1593	37	14.36	23	71.80	—
CV. (per cent)	24.16	16.31	—	—	—	—	—

to 37 days which was later than the original parent by 2 days. This indicated that all the mutants would mature in 62 to 64 days. The

100 seeds weight studies found that most of the mutants had larger seed size than the original parent. The mutant KKM 20-21-2-1-2 derived from the 20 kR had the highest 100 seeds weight with 26.33 gm. In comparison to the original parent this mutant had twice in seed size. The mutants treated at 40 kR, 60 kR and 80 kR had the average 100 seeds weight of 14.87, 14.38 and 14.36 grams respectively. However, they had larger seed size as compared to the original parent.

From the previous studies found that the number of pod/plant were related to pod size and seed size. The smaller pod and seed sizes would be higher the number of pod/plant. It was found that most of the mutants had higher number of pod per plant. The mutants treated at 60 kR and 80 kR had higher number of pod/plant as compared to the mutants treated at 20 kR and 40 kR. The mutant KKM 60-1-1-1 derived from the treatment at 60 kR got the highest number of pod per plant with 34 pods. Study on shelling percentage found that most of the mutants had high shelling percentage and did not different from the original parent. However, the mutant KKM 80-17-1-1-1 got the highest shelling percentage with 81 per cent.

Study on resistance to the diseases found that late maturity mutants tend to resistant to bacterial leaf blight disease, however late maturity mutants were attacked by insects severely, resulting lower seed yield. Most of the mutants had the diseases range from 0.5-2. Some of the mutants infected by Cercospora disease, the other were infected by bacterial leaf blight and other diseases. There was only one mutant (KKM 20-21-3-2-1) in the treatment at 20 kR, 4 mutants from 40 kR (KKM 40-3-2-1-1, KKM 40-3-3-1-1, KKM 40-5-2-1-1 and KKM 40-40-7-1-1), 4 mutants from 60 kR (KKM 60-10-1-1-1, KKM 60-6-2-1-2, KKM 60-16-1-1-1 and KKM 60-9-1-1-1-1) and 4 mutants from 80 kR (KKM 80-13-3-1-1, KKM 80-6-1-1-1, KKM 80-23-1-1-1 and KKM 80-9-1-1-1-1) were tolerant to the diseases. It was observed that, higher the rate of the treatments, higher the number of resistance mutants to the diseases were obtained.

Conclusion

From these yield trials some of the mutant lines were not stabled, some of the dark-red seed coat mutants were further segregated into black and red seed coats. The original parent was susceptible to bacterial leaf blight disease. There were 13 mutant lines tolerant to the diseases, while the remaining mutants had the diseases ranges from 5-40 per cent , resistance to the diseases were related to late maturity. Two mutants derived from the treatment at 60 kR (KKM 60-10-1-1-1 and KKM 60-12-2-1-1) got the highest seed yield more than 2 tons/ha. The mutant KKM 20-2-2-1-2 had the largest seed size (26.33 gm/100 seeds). The mutant KKM 60-16-1-1-1 got the highest number of pods/plant. While the mutant KKM 80-17-1-1-1 got the highest shelling percentages. It was found that high yielding mutants were associated to the number of pods/plant, 100 seeds weight, shelling percentages and resistant to the diseases.

In rainy season 1991, 25 mutant lines selected from all radiated treatments were tested for seed yield at KhonKaen University Experimental Farm. They were divided into two sets of yield trials. The first set consisted of 12 mutant lines and the second set had 13 mutant lines. They were planted and records on some of characteristics were made similar to the previous trials.

Mean seed yield of the first yield trial was 1288 kg/ha (Table 6.7). There were 5 mutants and 3 KKU varieties yielded higher than the average. The mutant KKM 60-6-1-1-1 got highest yield 1750 kg/ha, followed by KKM 60-9-2-1-2 and KKM 40-5-2-1-1 yielded 1524 and 1506 kg/ha respectively. In comparison to the original parent all the mutants yielded higher ranging from 6-109 per cent . There were four mutant lines (KKM 60-6-2-1-1, KKM 60-9-2-1-2, KKM 40-5-2-1-1 and KKM 60-6-2-1-2) resistant to bacterial leaf blight disease.

In the second yield trial, mean seed yield was 1294 kg/ha (Table 6.8). There were 5 mutants and one KKU variety yielded higher than an average. As compared to check there were eight mutants yielded higher. A mutant KKM 80-9-2-2-1 had highest yield 1463 kg/ha followed by mutant lines KKM 40-4-1-1-1 and KKM 20-1-1-1-1-4 yielded 1390 and 1381 kg/ha respectively. In comparison to the original parent, all the mutant lines yielded higher ranging from 25-98 per cent . Four mutants (KKM 80-9-2-2-1, KKM 40-4-1-1-1, KKM 40-5-2-1-2 and KKM 20-17-1-1-2 were resistant to bacterial leaf blight disease. It was found that higher yield were associated to number of pod per plant, 100 seeds weight, shelling percentage and resistant to bacterial leaf blight. Seven mutants were selected for further yield trials.

Table 6.7: Seed Yield and Certain Agronomic Characters of Cowpea Mutant Lines Grown at Khon Kaen During Rainy Season in 1991

Mutant Lines	*Yield kg/ha*	*Day to Flower*	*No. of Pod/ Pi*	*100 Seed wt. (g)*	*B-b (1-5)*	*Shel-ling (%)*	*D-m —*
KKU 305	1819A	36	22	16	R	82	63
KKM 60-6-2-1-1	1750AB	35	19	19	R	78	63
KKM 60-9-2-1-2	1524 ABC	36	14	17	R	77	63
KKM 40-5-2-1-1	1506 ABC	35	15	19	R	77	63
KKM 20-1-1-1-1	1444 BC	35	14	16	2.5	74	63
KKU264-1-3 (check)	1406 CD	36	14	18	2	75	65
KKU 402	1388 CDE	35	19	15	R	76	62
KKM 20-17-1-1-3	1375 CDE	35	18	16	1	73	63
KKM 60-6-1-2-2	1244 CDEF	34	16	15	2.5	67	61
KKM 80-9-2-2-1-2	1230 CDEF	34	19	16	2	73	60
KKM 60-6-2-1-2	1113 DEFG	32	22	14	R	69	56
KKM 60-16-2-1-1	1094 DEFG	33	13	16	2	81	61
KKM 20-2-3-2-1	1056 EFG	36	23	15	2	67	63
KKM 60-14-1-2-1	906 FG	35	20	17	2.5	72	62
KKM 60-6-1-1-1-3	008 G	32	17	15	2.5	66	57
Red cowper 6-1 U.S.	838 G	35	16	12	4	75	61
Mean	1288	35	18	16	1.4	74	62
CV. (per cent)	15.23	—	—	—	—	—	—

KKM = Khon Kaen Mutation

Letter 20, 40, 60 and 80 after KKM refer to treated rate at 20, 40, 60 and 80 kR.

B-b = Bacterial leaf blight disease., D-m = Day to maturity

During dry season in 1992, 12 cowpea mutant lines selected from 20, 40, 60 and 80 kR radiation treatments along with the original parents (Red cowpea 6-1 U.S).

KKU 402 and KKU 264-1-3 were planted by given KKU 402 as a check. They were tested for seed yield and resistant to bacterial leaf blight.

Table 6.8: Seed Yield (Kg/ha) and Certain Agronomic Characters of Cowpea Mutant Lines Grown at Khon Kaen During Rainy Season in 1991

Mutant Lines	*Yield kg/ha*	*No. of Pod/ Pt*	*100 Seed wt. (g)*	*B-b (1-5)*	*Shel-ling (%)*	*D-m —*
KKU 305	1650A	27	17	R	80	65
KKM 80-9-2-2-1	1463 AB	23	13	R	71	61
KKM 40-4-1-1-1	1390 ABC	20	21	R	78	63
KKM 20-1-1-1-4	1381 BC	17	16	1	72	59
KKM 40-5-2-1-2	1381 BC	17	20	R	76	63
KKM 60-1-1-1-1	1313 BCD	23	12	2.5	67	62
KKM 60-6-1-1-4	1288 BCD	21	13	2	76	59
KKM 60-6-1-1-6	1275 BCD	20	12	2	67	63
KKM 20-3-2-1-1	1250 BCDE	18	20	2	69	63
KKU 264-1-3 (check)	1200 BCDE	17	19	2.5	73	65
KKM 60-6-1-1-2	1175 CDEF	20	12	2.5	66	62
KKM 20-17-1-1-2	1150 CDEF	12	16	R	75	61
KKM 60-6-2-1-4	1044 DEF	17	12	2.5	67	60
KKM 60-6-2-1-3	1000 EFG	21	10	2.5	71	63
KKM 60-14-1-2-1	925 FG	18	18	2	74	62
RedCowpea6-1 U.S.	738 G	20	11	3.5	72	62
Mean	1294	19	15	1.7	72	62
CV. (per cent)	13.14	—	—	—	—	—

An average mean of seed yield was 2069 kg/ha. There were 7 mutant lines yielded higher than the check, but only one mutant line (KKM 40-5-2-1-1-1) yielded significantly difference from the check. In comparison to the original parent all the mutants yielded higher ranging from 7-32 per cent . Mutant line KKM 40-5-2-1-1-1 got the highest yield 2381 kg/ha follow by KKM 60-6-1-1-1-6 and KKM 60-6-2-1-1-1 yielded 2313 and 2244 kg/ha respectively, while the check KKU 402 and Red Cowpea 6-1 U.S. yielded 2044 and 1800 kg/ha. Studies on characters related to seed yield on KKM 40-5-2-1-1-1 found that this mutant had larger seed and highest shelling percentage, while number of pod/plant was slightly lower than an average. The mutant KKM 60-6-1-1-6 had number of pod per plant and number of seed per pod higher than an average, while 100 seeds weight and shelling percentage were lower than an average. The mutant line KKM 60-6-2-1-2 got lowest seed yield and had all the characters related to seed yield lower than an average. There was no disease occurred because of high temperature and low humidity at the time of flowering and 7 superior mutant lines were selected for planting in rainy season.

Average day to 50 per cent flowering was 50 days indicated that all of them were late flowering, because low temperature at the time of planting causing delayed of seed germination. An average number of pod per plant was 23 pods. There were 5 mutants had number of pod per plant higher than the average. It was found that all the mutants had number of pod per plant lower than the check and the original parent. Average number of seed per pod was 13 seeds and 4 mutants had number of seed per pod higher than the average. The mutant lines KKM 60-6-1-1-1-6-1, KKM 60-6-1-1-1-2 and KKM 20-17-1-1-1-2 had number of seed per pod higher than the other mutants. For 100 seeds weight 2 mutant lines (KKM 40-4-1-1-1 and KKM 40-5-2-1-1-1) had 100 seeds weight 20.3 and 20 gram, while the original parent had 12.7 gram. It was found that most of the mutants had larger seed size as compared to the original parent. For shelling percentage, mutant KKM 40-5-2-1-1-1 and KKM 40-4-1-1-1 had 81.5 and 81.4 per cent while the check and the original parent had 77.2 per cent and 75.8 per cent of shelling percentage. Average days to harvesting was 70 days indicated all the mutant lines were late maturity as compared to the original parent matured in 65 days. It was found that no bacterial leaf blight disease occurred during dry season planting.

Conclusion

It is concluded that mutant lines KKM 40-5-2-1-1-1 got the highest seed yield and higher than the original parent 32 per cent. It had higher number of pod per plant, 100 seeds weight and shelling percentage. There was no bacterial leaf blight appeared during planting. It was found that mutant derived from 40 kR radiated had larger seed and higher shelling percentage, while mutants derived from 20, 60 and 80 kR did not have those characters.

Table 6.9: Seed Yield and Certain Agronomic Characters of Cowpea Mutant Lines Grown at Khon Kaen During Rainy Season in 1992

Mutant Lines	*Yield kg/ha*	*DFF (50 %)*	*No. of Pod/ Pt*	*No. of Seed/ pod*	*100 Seed wt. (g)*	*Shel-ling (%)*	*D-m (Day)*
KKM 40-5-2-1-1	2381 A	51	24	12	20.0	81.5	72
KKM 60-6-1-1-1-2-1	2313 AB	48	27	15	13.0	70.0	70
KKM 60-6-2-1-1-1	2244 ABC	51	19	13	19.0	77.5	73
KKM 20-17-1-1-2	2213 ABC	50	21	14	16.8	75.1	72
KKM 60-6-1-1-1-2	2213 ABC	48	24	15	13.0	69.8	70
KKM 60-6-1-1-1-4	2156 ABCD	52	21	12	12.5	76.5	73
KKU 40-4-1-1-1	2075 ABCD	53	23	11	20.3	81.4	73
KKU 402-6 (check)	2044 BCD	52	35	12	17.5	77.2	73
KKM 80-9-2-2-1-2	1988 BCD	48	24	12	14.7	72.1	79
KKM 20-17-1-1-1-3	1963 CD	49	17	15	17.0	76.1	79
KKM 80-9-2-2-1-1	1960 CD	47	24	12	14.2	72.0	77
KKM 60-6-1-2-2	1956 CD	48	19	13	18.0	71.0	70
KKM 60-6-2-1-2	1938 CD	50	20	10	13.4	67.5	71
KKU 264-1-3-2 (B)	1813 D	52	22	11	22.5	75.1	76
Red Cowpea 6-1 U.S.	1800 D	47	28	13	12.7	75.8	65
Mean	2069	50	23	13	16.3	74.6	70
CV. (per cent)	11.25	—	—	—	—	—	—

During the rainy season in 1992. seven superior mutant lines were planted at Khon Kaen University experimental farm for seed yield and check for resistance to bacterial blight disease. They were tested in a Randomized Complete Block Designed in 4 replicates, 4 rows-plot, 5 meters long with 50 cm between rows and 20 cm between plants by planting KKU 264-1-3-2 as a check. Records were made on number of pod/plant, number of seed/pod, 100 seed weight, bacterial blight disease, shelling percentage and seed yield were made from 2 central rows.

Result and Discussions

Mean seed yield of all the cultivars was 1844 kg/ha (Table 6.10). There were 4 mutant lines yielded higher than the average. In comparison to the check and the original parent (Red Cowpea 6-1 U.S)., all the mutant lines yielded higher than both of them with 3 mutant lines yielded highly significant differences from the check. The mutants KKM 60-6-2-1-1, got the highest seed yield 2844 kg/ha followed by KKM 40-2-1-1-2 yielded 2794 kg./ha and KKM 40-4-1-1-1 yielded 2600 kg/ha, while the check and the original parent got the lowest yield with 1013 kg/ha. Studies on certain characters, such as number of pod per plant, number of seed per pod, 100 seed weight and shelling percentage found that, high yielding mutant lines were associated to higher number of pod per plant, larger seed size and higher shelling percentage. For bacterial blight disease studies found that the check and the original parent were susceptible to the disease, while the mutants KKM 60-6-2-1-1, KKM 40-5-2-1-1, KKM 40-4-1-1-1, KKM 60-6-2-1-2 and KKM 20-17-1-1-2 were resistant to the disease.

Table 6.10. Seed Yield (kg/ha) and Certain Agronomic Characters of Mutants Cowpea Grown at Khon Kaen During the Rainy Season in 1992.

Mutant Lines	*Yield kg/ha*	*No. of Pod/ Pt*	*No. of Seed/ Pod*	*100 Seed wt. (g)*	*Disease (1-5)*	*Shel-ling (%)*
KKM 60-6-2-1-1	2844	24	13	23	R	72
KKM 40-5-2-1-1	2794	28	14	21	R	74
KKM 40-4-1-1-1	2600	20	12	23	R	76
KKM 80-9-2-2-1	2244	22	12	16	1.5	66
KKM 20-17-1-1-2	1450	16	16	16	R	69
KKM 60-1-1-1-2	1350	18	12	13	1.5	61
KKM 60-6-2-1-2	1313	18	12	15	R	60
KKU 264-1-3-2(check)	1013	16	12	22	2.5	61
Red cowpea 6-1 U.S.	1013	20	12	14	4	66
Mean	1844	20	12.7	18	1.01	67
LSD (0.05)	356	—	—	—	—	—
C.V. (per cent)	13.22	—	—	—	—	—

Conclusion

It was found that 3 mutant lines (KKM 60-6-2-1-1-1, KKM 40-5-2-1-1-2 and KKM 40-4 1-1-1) got the highest yield. High yield was found to be associated to the number of pod per plant, 100 seed weight and shelling percentage. There were five mutant lines resistant to bacterial blight disease, while the original parent (Red Cowpea 6-1 U.S). and the check were susceptible to the disease. This disease might cause lower yield in the check and the original parent. Similarly, mutant lines resistant to this disease might be attributed to higher yield.

In 1993, 5 superior mutant lines were tested for advance yield trial. They were planted at Khon Kaen University Experimental farm, Ubol Research Institute Experimental farm and Mukdaharn Research Station Experimental farm by comparing the performance of mutant lines with the original parent (Red Cowpea 6-1 U.S). and the standard variety (KKU 305). The experimental design was a Randomized complete block design with 4 replicates, 4-rows plot, 5 meters long with 50 cm. between rows and 20 cm between plants. Combined analysis of variance over 3 locations were carried out for seed yield and certain characters.

Result and Discussions

Mean seed yield of all the mutant lines were 1,331 kg/ha (Table 6.11) with 3 mutants yielded higher than the check. In comparison to the original parent (Red Cowpea 6-1 U.S)., all the mutant lines yielded higher than the original parent, with 2 mutant lines (KKM 40-4-1-1-1 and KKM 60-6-2-1-1-1) yielded significantly differences from the check. The mutant KKM 40-4-1-1-1 got the highest seed yield 1,731 kg/ha followed by the mutants KKM 60-6-2-1-1-1 yielded 1688 kg/ha and KKM 40-5-2-1-1-1 yielded 1544 kg/ha, while the check and the original parent yielded 1,428 and 825 kg/ha, respectively.

Studies on yield components such as number of pod per plant, number of seed per pod, 100-seed weight and shelling percentage found that, the mutant lines which got higher seed yield had higher number of pod per plant, higher shelling percentages and larger seed size. For bacterial leaf blight disease studies, it was found that, a standard variety (KKU 264-1-3), a mutant KKM 60-6-2-1-2 and the original parent (Red Cowpea 6-1 U.S). were susceptible to the disease. The mutants KKM 40-4-1-1-1, KKM 60-6-2-1-1-1, KKM 40-5-2-1-1-2, KKM 80-9-2-2-1-2 and the check (KKU 305) were resistant to this disease.

Table 6.11: Combined Analysis of Seed Yield from 3 Locations and Certain Agronomic Characters of Cowpea Mutant Lines Grown at Khon Kaen, Mukdaharn, Ubol During Rainy Season in 1993

Mutant Lines	*Yield (kg/ha)*	*Pod/ Plant*	*Seed/ Pod*	*100 Seed wt (g)*	*Shelling (%)*	*B-d (1-5)*	*D-m —*	*Seed Color*
KKM 40-4-1-1-1	1731	11.9	11.9	20.8	77.8	R	69	Red
KKM 60-6-2-1-1	1688	13.6	11.2	18.6	75.3	R	69	Red
KKM 40-5-2-1-1	1544	12.2	11.5	19.6	74.0	R	70	Dark Red
KKU 305 (check)	1438	11.5	17.6	17.1	78.3	R	70	Black
KKM 80-9-2-2-1	1306	11.4	14.2	15.1	67.3	R	70	Red
KKU 264-1-3	1131	11.8	11.1	19.6	72.0	2	72	Black
KKM 60-6-2-1-2	969	10.4	13.1	14.6	61.1	2	68	Red
Red Cowpea 6-1US	825	10.7	11.3	11.5	69.0	3	65	Red
Mean	1331	11.7	12.8	17.1	71.9	0.88	69	—
CV.(per cent)	16.50	24.5	10.4	5.6	4.3	—	—	—
LSD (0.05)	178	0.61	1.56	0.48	1.55	—	—	—

Note: Rating on bacterial leaf blight disease (B-d) rating from 1-5

1 = lowest and 5 = highest disease infected

Conclusion

It was found that all the mutants yielded higher than the original parent ranging from 17-110 per cent . Three mutants yielded higher than a check ranging from 7-20 per cent , but only two mutants(KKM 40-4-1-1-1 and KKM 60-6-2-1-1-1) yielded significantly differences from the check. They had larger seed than the original parent ranging from 27-81 per cent . Higher seed yield were associated to higher number of pod per plant, 100 seed weight shelling percentage and resistant to bacterial leaf blight disease. Four mutant lines KKM 40-4-1-1-1, KKM 60-6-2-1-1-1, KKM 40-5-2-1-1-2, KKM 80-9-2-2-1-2 and the check (KKU 305) were resistant to this diseases, while the original parent, the standard variety KKU 264-1-3 and a mutant KKM 60-6-2-1-2 were susceptible to the diseases.

REFERENCES

IAEA. 1977. Induced mutation techniques in breeding seed propagated species. Manual on mutation breeding 3rd edition, p. 125-158.

Luadthong, S. and Thinnakorn Klomsa-ard, 1986-1990. Cowpea mutation breeding for resistance to bacterial blight (Xanthomonas vignicola Burk). Improving food and Agricultural production with Nuclear and related technology, Project abstracts in Plant mutation breeding, p. 31-32.

Luadthong, S. and Kongsak Wongthananan, 1987. Improvement of cowpea Production in Northeast Thailand. Contract Research of Northeast Crop Development Project Report. p. 78-121.

Luadthong, S. and Khajornwit Bhanyang-nao, 1992. Yield Trial of Cowpea Mutant lines, Targeted for Resistance to Bacterial Leaf Blight Xanthomonas vignicola (Burk). Fourth Nuclear Science and Technology Conference, p. 253-261.

Luadthong, S. 1993. Cowpea Mutation Breeding for Resistance to Bacterial Leaf Blight. Xanthomonas vignicola (Burk). Varietal Improvement of Cultivated Plant in Northeast Area of Thailand through Biotechnology, Faculty of Agriculture, Saga University, Japan, p. 39-41.

Chapter 7

Mutation Breeding: A Novel Technique for Quality Improvement of Winged Bean [*Psophocarpus tetragonolobus* (L.) DC.]

R.D. Dadke & V.S. Kothekar

Department of Botany, Dr. Babasaheb Ambedkar Marathwada University, Aurangabad – 431 004 (Maharashtra), India

ABSTRACT

Winged bean {*Psophocarpus tetragonolobus* (L.) DC.] has been described as a wonder legume for the tropics chiefly in view of its high protein content and a versatile utility potential. Through mutation breeding extensive efforts were made to accomplish the qualitative improvement of winged bean pertaining to its crucial biochemical features like the seed proteins and seed trypsin inhibitors. Excellent variability has been obtained in regard to such parameters and significant success has been gained in developing and characterising the low trypsin inhibitor

(TI) containing lines of winged been through mutation breeding. Such lines can be released as new varieties for the large-scale commercialisation of winged bean.

INTRODUCTION

Seed legumes, "The poor man's meat", are universally accepted as key sources of proteins and comprise major parts of human diet in several developing countries. The wealthy nations have high levels of nutrition and face little problem in supplying all their citizens with quality food. In the developing world, however, the scenario is quite different. More than one billion people live in utter poverty and about half of that number suffers from serious protein deficiency and malnutrition. At the same time there is a trend of partial replacement of animal food by legume proteins from the point of view of nutrition and socio-economic impact. If we take these two factors into consideration, for increasing the production of legume seeds appreciably, the legume agriculture needs to be given substantial push and input support. The current situation indicates that the overall production of seed legumes is very less as compared to the cereal production. This has happened due to the excessive emphasis on the cereal production through the so-called 'Green Revolution'. In the light of this, a similar revolution in the production of pulse legumes has been considered necessary which would quickly fulfil the protein requirement of the world in near future.

Thousands of legume species are known to mankind but our conventional agriculture practices have kept on revolving around only chosen few species. Now-a-days less than 20 species of legumes are domesticated for various utility purposes. The major ones are the chickpea, pigeon pea, peanut, cowpea, moth bean, green gram, soybean etc. Apart from these cultivated species, there exist many under-exploited legume species. Such neglected species may follow the developmental course of the established commercial legume varieties provided; proper scientific attention and research efforts are directed towards them.

One such unexploited and under-exploited seed legume is the winged bean, *Psophocarpus tetragonolobus* (L). DC. The winged bean has successfully demonstrated an exceptionally fast rate of dispersal, development and recognition as a new food crop throughout the tropical region of the world. Masefield (1973) was the first person who could highlight the potential utility of this plant. The worldwide interest in this plant got triggered subsequently through the publication entitled, "Winged Bean–A High Protein Crop For The Tropics" brought out by the National Academy of Sciences, US (NAS, 1975).

A Versatile Legume

The winged bean looks like a pole bean. It is an annual/ perennial twinner and attains a height of four meters on proper staking. It has the ability to nodulate with wide spectrum of rhizobia. It is widely adaptable to diverse soil types and not very demanding as regards the soil requirements. It can be grown in the tropical and subtropical conditions.

A Protein Rich Food

The nutritional value of the winged bean is mainly due to its mature seeds. The seeds contain high amount of protein (29 per cent –42 per cent) and are thus comparable to those of soybean (NAS 1981). The amino acid profile in the winged bean is quite similar to that of soybean (Claydon 1978, 1979, Haq 1982). It is due to this similitude, various researchers are now hailing the winged bean as, "A tropical soybean". Besides the protein, the winged bean seeds contain about 20 per cent edible oil (Claydon 1978, 1979). The fatty acid profile of winged bean has demonstrated high amount of unsaturated fatty acids and shows a good analogy with groundnut oil (Cerney *et al.*, 1971). The most interesting feature of the winged bean in some varieties is the tuber. The tuber contains significantly higher amount of proteins (20 per cent) than any other tuber crops (NAS 1981). Also the other plant parts can be consumed or profitably utilised. In view of this, the winged bean has been described as "Supermarket on a stalk".

Despite possessing several good qualities and nutritional potential, so far the winged bean has remained obscure and unfamiliar. This has happened due to some undesirable features possessed by the plant. Some such features include (1) the long duration of the crop besides late maturity, 2) the twiny/climbing habit of the plant with a strong need for staking and 3) the presence of anti-nutritional factors in various plant organs.

Role of Induced Mutation

Today, the genetic improvement of plant is undertaken either by using conventional plant breeding or the modern methodological approaches like the biotechnology. The third angle for plant improvement is the mutation breeding. All such techniques have their own merits and demerits. Plant breeding is still recognized as a sure approach for architecturing the crop variety. But on occasions it so happens that a particular desirable character in which the plant breeder is interested does not become available in the gene pool of cultivated/wild varieties. Now-a-days the biotechnology has opened up a new vistas in agriculture and has offered innumerable possibilities for engineering the genetic make up of the plant in a desired form. Despite this strong merit, the environmentalists consider it as unsafe technique and have their strong stand against the release of genetically modified plants. In addition to this the biotechnology comprises a highly sophisticated and costly technique, which requires lot of infrastructure, high inputs besides specialized expertise.

In the backdrop of this, the mutation breeding has been considered as one of the gizmo of the atomic age and a much promising breeding short-cut which could be beneficially utilized for tailoring the desired varieties of different crop plants. The major advantage of the use of induced mutations is the possibility to correct one or few negative characters of a cultivar, without changing the major parts of its total genetic set up. Mutation breeding has now become a novel and eco-friendly but useful and in some situations even an indispensable tool which can be employed by the plant breeders.

Aims of Mutation Breeding in Winged Bean

We know that winged bean is a climber and its natural germplasm lines do not carry any genes for dwarfism. The development of determinate varieties with a self-standing bushy architecture has assumed importance for taking full advantage of winged bean as a cultivated legume crop. Several research workers like Khan and Brock (1975), Khan and Stephenson (1977), Kesavan and Khan (1978) and Karikari (1981) achieved moderate success in developing bushy mutant of winged bean through gamma radiation. This activity revealed the possibility of obtaining a self-standing dwarf plant type through induced mutations.

Keeping this end in view we initiated mutation breeding programme in winged bean for achieving its genetic improvement under our local conditions. The pertinent study, spread over more than a decade, resulted in the development of several interesting true breeding macro-mutants induced through the treatments of gamma rays, EMS and NEU in winged bean.

Biochemistry of Winged Bean Mutants

The importance of winged bean as a source of proteins has already been recognized but few anti-nutritional components like trypsin inhibitors are limiting the wide scale use of winged bean. The trypsin inhibitors are protein in nature and rich in sulphur containing amino acids. The physiological functions of trypsin inhibitors are either regulatory *i.e.* controlling the activity of endogenous proteases in storage organs before/during germination or of a defense role against the digestive proteases of the animals and microbes (Ryan 1981).

In the present study, we have biochemically screened the different true breeding winged bean mutants in regard to the status of their protein content and the quality parameter like trypsin inhibitors prevailing in them. Also we shall discuss the correlation of trypsin inhibitory activity, their isoinhibitors detected through electrophoretic technique and the overall protein quality of the seeds of winged bean mutants.

Materials and Methods

Material

Sixteen true breeding M10 mutant lines of variety iiHP Sel-21 of winged bean were taken for the biochemical studies. The list of mutants is given in Table 7.1.

Crude Protein Estimation

This was estimated by Microkjeldahl method.

Protein Extraction

The fine seed powder was suspended in extraction buffer (50mM Tris-Cl, pH 8 containing 1mM CaC12) for 2 h. at room temperature in centrifuge tube. After centrifugation at 14500 r.p.m. for 30 minutes, the clear supernatant was saved as a crude extract.

Table 7.1: The List of Mutants of Winged Bean

1.	*Xantha*	10.	Linear leaflet/flat pod
2.	Anthostem	11.	Dark green/flat pod
3.	Wingless pod	12.	Linear leaflet
4.	Long pod I	13.	Triangular leaflet
5.	Lond pod II	14.	Large leaf
6.	Anthobud	15.	Large leaf/stiff stem
7.	Early maturing	16.	Dwarf
8.	Flat pod	17.	iiHP Sel 21 (control)
9.	Long flat pod		

Protein Estimation

Protein estimation was carried out by the Biuret method (Layne 1957).

Determination of Trypsin Inhibitor Activity

Trypsin inhibitor (TI) activity was assayed according to the method of Erlanger *et al.* (1961) using N-benzoyl-DL-arginine-p-nitroanilide hydrochloride (BAPNA) as a chromogenic substrate. The TI content was expressed as trypsin inhibitory unit per g of seed meal.

Electrophoresis

Non-denaturing discontinuous polyacrylamide gel electrophoresis was performed in a Hoeffer SE-400 vertical slab gel apparatus with 1 mm thickness. The gel was prepared according to Davis (1964) system. Equal protein samples (40 ug) of seed proteins were loaded. Electrophoreesis was carried out at constant voltage. After electrophoresis the gel was fixed in 10 per cent TCA and subsequently stained with Coomassie brilliant blue R-250 dye.

Electrophoretic Detection of TI

After non-denaturing discontinuous electrophoresis, the unstained gel was used for X-ray film contact method for detection of trypsin isoinhibitor bands, according to the technique developed by Pichare and Kachole (1994).

Results and Discussion

Protein Content

The crude protein content ranged from 26.45 per cent to 33.32 per cent in the seeds of different mutants (Table 7.2). Almost all mutants showed enhancement in the crude protein content over the control (26.45 per cent). The highest protein content could be noticed in the *Xantha* (33.32) and anthostem (33.16 per cent) mutants.

Table 7.2: Biochemical Parameters of Winged Bean Seeds

Sl.No.	*Name of the Mutant*	*Crude Protein Content (%)*	*Extractable Protein Content (mg/g)*	*Trypsin Inhibitor Content (TIU/g)**
1.	Control	26.45	161.01	9806.45
2.	Xantha	33.32	197.36	19013.15
3.	Anthostem	32.16	134.61	16217.94
4.	Wingless	29.30	137.41	7019.86
5.	Long pod I	28.79	142.40	10949.36
6.	Linear leaflet	31.12	140.12	12675.15
7.	Anthobud	29.15	158.22	6392.40
8.	Early maturing	31.38	154.80	4193.54
9.	Linear leaflet/flat pod	31.12	109.27	7549.66
10.	Dark green/flat pod	27.49	169.93	12483.66
11.	Large leaf/stiff stem	30.86	164.01	8152.86
12.	Long pod II	30.34	166.66	8846.15
13.	Large leaf	27.59	162.25	8079.47
14.	Triangular leaf	26.45	187.89	10509.55
15.	Flat pod	31.23	105.96	5099.33
16.	Long flat pod	29.05	168.99	9610.38
17.	Dwarf	31.23	151.33	8606.31

* Mean of three values

While the extractable seed protein content in the control (Table 7.2, Fig. 7.1) was 161.01 mg/g, again, the *Xantha* (197.36 mg/g). mutant revealed enhancement as regards that feature over the control. In all other mutants, the extractable protein content ranged from 134.61 mg/g. to 169.93 mg/g. except for the flat pod mutant (Fig. 7.2), which exhibited the lowest value of 105.96 mg/g.

Earlier several researchers have reported altered protein content through induced mutations (Haq *et al.*, 1970, Gaul *et al.*, 1973, Walther *et al.*, 1975, Rabson and Bhatia 1975, Reddy 1977) in different crop plants. Interestingly, the crude protein content and extractable protein content of *Xantha* (Fig. 7.3), a chlorophyll mutant, were found to be higher than the control. Generally, the chlorophyll mutants like, *Xantha, Chlorina, Albina* are the lethal mutations (Gustafsson 1938) and do not survive. But here in case of winged bean, we have noticed a viable true breeding M10 *xantha* mutant. Even we have incorporated this *xantha* mutant in intermutant hybridization programme in order to know its genetics. To our expectation, it was found to be recessive in nature, true breeding and segregating according to the Mendelian principle. The further studies in this regard are in progress at our end.

TI Content

The trypsin inhibitor content in the two mutants (Table 7.2), namely, early maturing (4193.54 TIU/g meal) and flat pod (5099.33 TIU/g meal) exhibited reduced values in contrast to the control (9806.45 TIU/g meal). The 40 per cent to 60 per cent reduction in the antitryptic activity in these two mutants, clearly indicates the use of induced mutations in enhancing the nutritional quality of the seed protein.

To date, very few reports are available in regard to the role of induced mutation for altering the TI activity in plants Kothekar *et al.* (1996), Khandelwal (1996) and Dadke (1999) reported the low TI mutants of winged bean. Harsulkar (1994) observed the altered TI activity in varied barley mutants.

Electrophoretic Detection of TI (Figs. 7.4 and 7.5)

The electrophoretic trypsin inhibitor profile of winged bean exhibited a maximum of seven bands of isoinhibitors in control and antho-bud mutant. The isoinhibitor band "C" was missing in large leaf/stiff stem mutant while the isoinhibitor bands "D" and "E" were feeble in the same mutant. The triangular leaflet mutant showed the absence of isoinhibitor band "D". The information available as regards the use of induced mutations in connection with trypsin inhibitor is next to negligible. Kothekar *et al.* (1996) reported altered

Fig. 7.1: The Plant Habit of W.B. Plane

Fig. 7.2: The Flat Pod Mutant

Fig. 7.3: The Xantha Mutant

Fig. 7.4: The Seed Trypsin Isoinhibitor Profile of Winged Bean mutants

Lane: C–Control, 1–Wingless, 2–Dark green flat pod, 3–Long flat pod, 4–Flat pod, 5–Linear leaflet, 6–Triangular leaflet, 7–long pod I, 8–Long pod II, 9–*Xantha*

Fig. 7.5: Seed Trypsin Isoinhibitor Profile of Winged Bean Mutants

Lane: C–Control, 10–Linear leaflet/flat pod, 11–Dwafr, 12–Large leaf, 13–Linear leaflet/stiff stem, 14–Antho bud, 15–Anthostem, 16–Early maturing

TI profile in winged bean mutants through the approach of induced mutagenesis.

Induced Mutation and Nutritional Aspect

All legume seed proteins are relatively low in sulphur containing amino acids (Rockland and Radke, 1981) and the presence of antinutritional compounds like trypsin inhibitors are further limiting the use of seed legumes from the nutritional angle. In view of this, consumption of legume proteins alongwith the cereal proteins is considered advisable, since cereals are rich in sulphur containing amino acids. Another constraint as regards the use of legume protein is the trypsin inhibitor mainly on account of its adverse dietary effects (Liener and Kakde 1969; Krogdahl and Holm 1981 and Higuchi *et al.*, 1983). For enhancing the protein digestibility, proper cooking/some sort of heat treatment of the seed legumes is always felt necessary (Liener and Kakade 1969).

Once inactivated, the trypsin inhibitors may even play a positive nutritional role due to their high content of sulphur containing amino acids relative to the majority of the seed proteins (Ryan 1990).

With this understanding, it could be argued that the approach of induced mutation is very useful for altering TI activity and TI quantity so that one can overcome the adverse dietary effects of the trypsin inhibitors by avoiding long cooking and heat treatment. Thus legume seeds could be made suitable even for raw consumption.

Our results clearly indicate that mutation can be induced in the inhibitory sites of the trypsin inhibitor. This was substantiated by the fact that no gross change in the band patterns of trypsin isoinhibitors of winged bean mutants could become noticeable.

Although the trypsin inhibitor content has been found to be appreciably reduced in few mutants, their protein contents, however, have remained high as compared to control. In this way, the dual advantage embodied by the inactivated trypsin inhibitors besides high protein content could be derived for the amelioration of seed protein quality of winged bean achieved through the approach of induced mutation.

From the foregoing, it could be inferred that the development of the mutants carrying lowered levels of trypsin inhibitors would be quite relevant in winged bean improvement as several parts of the plant are traditionally eaten in the green and uncooked form. It is further felt that considerable scope exists for incorporating such promising mutant types in conventional breeding programme for evolving quality recombinant lines of winged bean.

Acknowledgements

The authors acknowledge with thanks the financial support received from the Department of Atomic Energy, Government of India for undertaking the major part of the present work. They thank Dr. S.E. Pawar and Dr. T. Gopalakrishna from B.A.R.C., Mumbai, for useful discussion and encouragement.

REFERENCES

Cerney, K., Kordy, M., Pospisil, F., Suabensky, O. and Zajic, B. 1971. Nutritive value of winged bean (*Psophocarpus tetragonolobus* (L).DC). Br. J. Nutr., 29: 293-298.

Claydon, A. 1978. Winged bean–a food with many uses. Plant foods for man, 2(2): 203-224.

Claydon, A. 1979. The use of legumes as sources of edible oil (with special reference to winged bean) In: Legumes in the Tropics, Kualalumpur (Proc. Symp). 473-477.

Dadke, R.D. 1999. Characterisation of mutants and hybrids of winged bean *Psophocarpus tetragonolobus* (L). DC. Ph.D. thesis, Dr. B.A. Marathwada University, Aurangabad, India.

Davis, B.J. 1964. Disc electrophoresis II methods and application to human serum, Ann. N.Y. Acad. Sci. 122: 404-429.

Erlanger, B.F., Kokowsky, N. and Cohen, W. 1961. The preparation and properties of two new chromogenic substrates of trypsin, Arch. Biochem. Biophys. 95: 271-278.

Gaul, H., Ulonska, E., Lund, V. and Walther, H 1973. Studies on selection of high protein and lysine content in barley mutants. Symp, "Nuclear techniques for seed protein improvement", IAEA, Vienna: 209-215.

Gustafsson, A. 1938. Studies in the genetic basis of chlorophyll formation and the mechanism of induced mutations. Lund Univ. Arsskr. N.F. Adv. 5.

Haq, M.S., Ali, S.M., Maniruzuman, A.F.M., Mansur, A. and Islam, R. 1970. Breeding for earliness, high yield and disease resistance in rice by means of induced mutations. From "Rice Breeding with Induced Mutations", IAEA, Vienna: 77-78.

Haq, N. 1982. Germplasm resources, breeding and genetics of the winged bean. Z. Pflanzensuchtg. 88: 1-12.

Harsulkar, A.M. 1994. Studies on mutagenic effects of pesticides in barley. Ph.D. thesis, Dr. B.A. Marathwada University, Aurangabad, India.

Higuchi, M.M., Suga and Iwai, K. 1983. Participation in biological effect of raw winged bean seeds on rats. Agric. Biol. Chem. 47: 879.

Karikari, S.K. 1981. The effect of seed irradiation on plant characteristics and yield of winged bean *Psophocarpus tetragonolobus* (L). DC. Paper presented at 2nd International seminar on winged bean held at Colombo, Jan: 18-23.

Kesvan, V. and Khan, T.N. 1978. Induced mutation in winged bean. Effects of gamma rays and ethyl methanesulphonate. The winged bean 1st International symposium on developing the potentialities of winged bean, 1978, Manila, Philippines, 105-109.

Khan, T.N. and Stephenson, R.A. 1977. Genetic resources in winged bean and their utilization. International Congress of the society of the 4th advancement of breeding researches in Asia and Oceania (SABRO) in association with Australian plant breeding conference, Canberra, Australia, Feb: 19-25.

Khan, T.N. and Brock, R.D. 1975. Mutation breeding in winged bean. In: The regional seminar on the use of induced mutations in improvement of grain legume production in South-East Asia, Colombo, Sri Lanka, Dec. 8-12.

Khandelwal, A.R. 1996. Cytological, biochemical and mutational studies in *Psophocarpus tetragonolobus* (L). DC. Ph.D. thesis Dr. B.A. Marathwada University, Aurangabad, India.

Kothekar, V.S., Harsulkar, A.M. and Khandelwal, A.R. 1996. Low trypsin and chymotrypsin inhibitor mutants in winged bean *Psophocarpus tetragonolobus* (L). DC. J. Sci. Food Agric. 71: 137-140.

Krogdahl, A. and Holm, H. 1981. Soybean proteinase inhibitors and human proteolytic enzymes. Selective inactivation of inhibitors by treatment with human gastric juice. J. Nutr. 111: 2045-2051.

Layne, E. 1957. Spectrophotometric and turbidimetric methods for measuring proteins. In: Methods of enzymology, Edn. S.P. Colowick and N.O. Kaplan., Academic Press, New York, Vol. 3: 2252.

Liener, *I.E.* and Kakade, M.L. 1969. Protease inhibitors in "Toxic constituents in plant foodstuffs", Ed. Liener, *I.E.*, P-7, Academic Press New York.

Masefield, G.B. 1973. *Psophocarpus tetragonolobus* a crop with future? Field crop Abstract, 26: 157-160.

NAS 1975. The winged bean–A High Protein Crop of the Tropics, National Academy of Sciences, U.S. Govt. Printing Off. Washington D.C. Edn.-I.

NAS 1981. The winged bean–A High Protein Crop of the Tropics. National Academy Of Sciences, U.S. Govt. Printing Off. Washington D.C. Edn-II.

Pichare, M.M. and Kachole, M.S. 1994. Detection of electrophoretically separated protease inhibitors using X-ray film. J. Biochem. Biophys. Methods, 28: 260-264.

Rabson, R. and Bhatia, C.R. 1975. In Proc. II Int. Winter wheat Conf. Zagreb, Yugoslovia: 411.

Reddy, G.M. 1977. EMS induced grain shape mutations in some local cultivars of rice. II Riso. 26: 187-192.

Rockland, L.B. and Radke, T.M. 1981. Legume protein quality. Food Technol. 28: 79-82.

Ryan, C.A. 1981. Proteinase inhibitors in: The biochemistry of plants: Proteins and nucleic acids, Vol.6, Edn. Marcus, A., Academic Press, New York, 351.

Ryan, C.A. 1990. Protease inhibitors in plants: Genes for improving defenses against insects and pathogens. Ann. Rew. Phytopathol. 28: 425-449.

Walther, H., Gaul, H., Ulonska, E. and Seibold, K.H. 1975. Variation and selection of protein and lysine mutants in spring barley, from, "Breeding for Seed Protein Improvement Using Nuclear Techniques". IAEA, Vienna: 79-89.

Chapter 8

Quantitative and Qualitative Improvement in *Brassica* Oil Crops through Induced Mutation Technique in Bangladesh

M.L. Das & A. Rahman

Bangladesh Institute of Nuclear Agriculture, P.O.Box 4, Mymensingh, Bangladesh

ABSTRACT

Bangladesh Institute of Nuclear Agriculture by using induced mutation technique has developed two yellow seeded varieties, Safal and Agrani sharisha of *Brassica campestris*, and two brown seeded varieties, BINA sharisha-3 and BINA sharisha-4 of *Brassica napus*. All the four new mutant varieties outyielded the mother and check. Improvement in oil content and fatty acid composition was also obtained in all the mutant varieties. Further, Safal and Agrani sharisha were more tolerant to *Alternaria* blight compared to other varieties under field conditions. BINA sharisha-3 and BINA sharisha-4 were obviously two to three weeks early maturing compared to mother and check. The highest mutation frequency was observed in the progenies at 700 Gy followed by 900 Gy.

Maximum segregation was found in M_2 generation which provided a good scope for selection of desirable mutants. True breeding mutants were evaluated from the segregating population in M_3 and M_4 generations. However, the National Seed Board of Bangladesh approved all the above four mutant varieties for commercial cultivation in the farmers field all over the country. Cultivation of these two new varieties has already began in the farmers field.

INTRODUCTION

Mustard (*Brassica* spp). is a major edible oil yielding crop which ranks first among the oil crops grown in Bangladesh. Widely grown cultivars of *B. campestris* are low yielding (720 kg/ha) and highly susceptible to *Alternaria* blight disease. On the other hand, the cultivars of *B. napus*, though physiologically more productive (Tsunoda 1980), take more than 100 days to mature, plants are of spreading type and often lodge (Das and Rahman 1994, Das *et al.*, 1995). Long maturity period makes the crop unsuitable to grow in the existing rice based cropping pattern of the country. Keeping these constraints in mind, varietal improvement in *B. campestris* and *B. napus* using induced mutation technique was undertaken at the Bangladesh Institute of Nuclear Agriculture during 1980s. The technique was chosen for the present work since it is widely used with success for crop improvement in many countries of the world (Panda 1981, Micke *et al.*, 1987, Ashri 1988, Robbelen 1991, Novak and Brunner 1992, Rahman *et al.*, 1992).

The first objective was to evolve high seed yield potential varieties of *B. campestris* with resistance/tolerance to *Alternaria* blight and the second objective was to develop early maturing varieties of *B. napus* that could be suitably grown in the rice based cropping pattern of the country. This paper describes the successes in evolution of four improved varieties of *B. campestris* and *B. napus* by using induced mutation technique.

Material and Methods

One thousand and five hundred dry seeds per dose of a yellow seeded cv. YS-52 of *B. campestris* were exposed to 500, 600, 700, 800 and 900 Gy of gamma rays from a ^{60}Co source in 1981. Irradiation was carried out at a constant dose rate of 0.58 Gy per minute. Similarly, 1500 seeds per concentration were also treated with 0.25,

0.50, 0.75 and 1.00 per cent of EMS following the standard procedures as suggested by Anon. (1977). All the irradiated/treated seeds were sown direct in the field following randomized complete block design. At maturity, first formed 10 pods from each of the surviving plants were harvested to raise M_2 in plant progeny rows. Selection of micro–and macro-mutants was primarily done in M_2 generation. Further selections of true breeding elite mutants were made from the segregating population in M_3 and M_4 generations. Normal looking plants were selected which had higher number of pods/plant with resistance/tolerance to *Alternaria* blight under field conditions. Performances of the selected mutants compared to the existing varieties under cultivation (check) were evaluated at different agro-ecological zones of the country. All the experiments were laid out in randomized complete block design with four replications. Unit plot size was 4m × 5m having 20 cm spacing between rows. Recommended cultural practices (fertilizers, weeding, irrigation, etc). were followed to ensure normal growth of plants. Reaction of *Alternaria* blight disease was assessed by growing the mutants, parents and susceptible variety side by side.

Further, in order to induce earliness in *B. napus*, four different doses (700, 900, 1000 and 1100 Gy) of gamma rays were exposed to 1500 dry seeds per dose of Nap-3 (Mother) in 1989. All other procedures, such as experimental design, cultural practices, selection and evaluation of mutants, etc. were same as described above.

Agronomic data were collected from all the experiments carried out in different years following the standard techniques and subjected to appropriate statistical analyses. The mean values of different characters were adjudged by the Duncan's New Multiple Range Test using the formula as suggested by Steel and Torrie (1960). Oil content (per cent) of seeds and fatty acid profiles were determined by NMR and gas chromatography (GLC, Perkin Elmer 8410), respectively.

Results and Discussion

Improvement Made in a cv. YS-52 of *B. campestris*

Both gamma rays and EMS were potentially effective for creating variabilities in different characters of YS-52 (Mother). The analysis of variance showed that different treatment doses and concentrations produced significant variation in germination of seeds, plant height, number of branches/plant, number of pods/plant, number of seeds/pod, 1000-seed weight and seed yield/plant as published elsewhere (Rahman *et al.*, 1988). Segregation was found in different morphological characters in M_2 generation (Table 8.1). The highest frequency of morphological variation for all the characters was found at 700 Gy followed by 900 Gy of gamma rays while in EMS, 1.00 per cent concentration showed the highest frequency of morphological mutation followed by 0.75 per cent concentration. Variation in number of chambers (2, 2 + 3, 2 + 4, 2 + 3 + 4) of pods in the same plant was found against the 4 chambers' pod in the mother. Desirable mutants were not scored from EMS treatments in M_3 generation. Eight resistant and 46 tolerant mutants against *Alternaria* blight were isolated from the M_2 population at 700 and 900 Gy. However, only 11 mutants bred true in M_3 generation. Of them three mutants (selection No. BINA-1, BINA-2 and BINA-3) produced significantly higher number of pods/plant which outyielded the mother as reported elsewhere (Rahman *et al.*, 1988). These three mutants also showed tolerance to *Alternaria* blight under field conditions. However, yield trials were carried out with the best two mutants, BINA-1 and BINA-3 (later named as Safal and Agrani sharisha, respectively) and check during 1984-85 to 1989-90 (six years) at seven agro-ecological zones of the country. The result showed that the mean performances of both the mutant varieties were superior to all the check (Table 8.2). It was found from the six years' trials that Agrani sharisha significantly produced the highest mean seed yield (1.70 tons/ha) followed by Safal sharisha (1.68 tons/ha). The existing cultivated varieties, Sampad, Sonali and Tori-7 gave 1.38, 1.41 and 1.00 tons/ha, respectively. YS-52 (Mother) produced 1.30 tons/ha. In 1989-90 yield trials conducted in the farmers' field of seven agro-ecological zones also showed the superior performance of the mutant varieties over the check.

The Department of Agricultural Extension (DAE) has conducted 142 demonstration trials with these two mutant varieties in the farmers' field at different agro-ecological zones during 1993-94 to 1996-97 (four years). Safal and Agrani sharisha maintained their superior performance at the onfarm levels which, on an average, produced the highest seed yield of 1.31 and 1.28 tons/ha, respectively (Table 8.3). These two mutant varieties gave 21.1 and 17.2 per cent higher seed yield than the existing high yielding varieties, Sampad and Sonali sharisha.

Table 8.1: Segregation in Different Morphological Characters in the Mutagen Rreated M_2 Population of *Brassica campestris* var. Yellow Sarson

Mutagen in Different Doses/ Concentrations	*No. of Progeny Studied*	*Number of Progeny Segregated for Different Morphological Characters*				
		Chlorophyll Mutants	*Leaves*		*Flowers*	
		(Albina)	*Wax Deficient*	*Virescent*	*Closed*	*Semi Closed*
YS-52 (Mother)	1000	—	—	—		—
Gamma rays (Gy)						
500	882	6	—	—	—	—
600	600	3	—	—	3	3
700	1576	22	11	2	11	21
900	450	8	—	—	3	1
EMS (%)						
0.25	1355	—	—	—	—	—
0.50	1411	3	—	—	—	—
0.75	335	7	—	—	3	2
1.00	321	7	—	—	11	7

Mutagen, Doses/ Concn.	*Type of Plants*				*Number of Chamber in Pods*				*Reaction to Alternaria*		*To-tal*
	Ste-rile	*Bu-shy*	*Dw-arf*	*Ear-ly*	*2*	*2+3*	*2+4*	*2+3+4*	*R*	*T*	
YS-52	—	—	—	—	—	—	—	—	—	—	—
Gamma rays (Gy)											
500 Gy	—	—	—	—	—	—	—	—	—	—	6
600 Gy	—	—	—	—	—	—	—	—	—	—	9
700 Gy	20	3	14	2	183	2	25	1	8	46	371
900 Gy	11	—	2	—	122	—	2	1	—	—	150
EMS (%)											
0.25	—	—	—	—	—	—	1	—	—	—	1
0.50	—	—	—	—	—	—	—	—	—	—	3
0.75	3	3	2	6	1	—	4	—	—	—	31
1.00	11	8	2	—	—	—	1	—	—	—	47

* R = Resistant, T = Tolerant to *Alternaria* blight.

Source: Rahman *et al.* (1988)

Table 8.2: Seed Yield of Two Mutant Varieties Compared to Check During 1984-85 to 1989-90 in 7 Different Agro-Ecological Zones (AEZ) of Bangladesh

Variety	*Seed Yield (tons/ha)*							
	**PYT at 3 AEZ in 1984-85*	*SYT at 4 AEZ in 1985-86*	*AYT at 4 AEZ in 1986-87*	*RYT at 4 AEZ in 1987-88*	*RYT at 3 AEZ in 1988-89*	*ZYT at 3 AEZ in 1989-90*	*Mean*	*Farmers' field at 7 AEZ in 1989-90*
Mutant Varieties								
Safal	1.45a	1.85a	1.75a	1.82a	1.48a	1.73b	1.68a	1.40a
Agrani	1.43a	1.80a	1.70a	1.81a	1.49a	1.97a	1.70a	1.37a
Check:								
Sampad	1.28b	1.40c	1.33c	1.40c	1.30b	1.55c	1.38b	1.16b
Sonali	1.24b	1.56b	1.43b	1.52b	1.20c	1.53c	1.41b	1.35a
Tori-7	0.89e	1.09d	1.04d	-	-	-	1.00d	0.87c
YS-52	1.18d	1.43c	1.49b	1.31d	1.09d	-	1.30c	-

Values having common letter (s) in each column do not differ significantly at 5 per cent level.

* PYT = Preliminary Yield Trial, SYT = Secondary Yield Trial, AYT = Advanced Yield Trial, RYT = Regional Yield Trial, ZYT = Zonal Yield Trial.Table 8.3

Table 8.3: Seed Yield of Safal and Agrani Sharisha in the Famers' Field During 1993-94 to 1996-97

Year	*Number of*		*Mean Seed Yield (tons/ha)*				*% Increased Over Check*	
	Dis-tricts	*Far-mers Field*	*Safal Sha-risha*	*Agrani Sha-risha*	*Sampad (Ch-eck)*	*Sonali (Ch-eck)*	*Safal Sha-risha*	*Agrani Sha-risha*
1993-94	6	29	1.43a	1.32a	1.11b	—	32.9	18.0
1994-95	4	35	1.19a	1.20a	—	1.15a	18.4	19.9
1995-96	12	50	1.12a	1.11a	—	1.04a	10.6	9.3
1996-97	6	28	1.50a	1.47a	—	1.28b	22.3	21.7
Average of 4 years			1.31	1.28	1.11	1.16	21.1	17.2

Values having common letter (s) in each year do not differ significantly at 5 per cent level.

Source: Department of Agricultural Extension, Govt. of Bangladesh.

Reaction of mutant varieties and check to *Alternaria* blight was assessed at five different agro-ecological zones. Disease symptoms appeared first at 30 days after sowing which was increasing with the increase of age of plants up to maturity. Generally, leaves and pods were affected. However, varying degrees of infection were found in the mutant varieties and check under study. Safal and Agrani sharisha along with Sonali were observed to be tolerant to *Alternaria* blight in field conditions compared to Sampad and YS-52 which showed susceptible to this disease (Table 8.4).

Table 8.4: Reaction of Safal and Agrani Sharisha and Check to *Alternaria* Blight in Field Conditions at 5 Different Agro-Ecological Zones of Bangladesh

Variety/Check	*% Leaf and Pod Area Affected with Alternaria Blight*						*Disease Reaction*
	L-1	*L-2*	*L-3*	*L-4*	*L-5*	*Average*	
Mutant Varieties							
Safal	23	31	29	41	27	30.2	T
Agrani	35	—	31	29	30	31.3	T
Check							
Sonali	37	42	33	35	31	35.6	T
Sampad	49	42	43	39	41	42.8	S
YS-52 (Mother)	51	42	43	46	44	45.2	S

L = Location in agro-ecological zone.

T = Tolerant (when 26–40 per cent leaf and pod area were affected)

S = Susceptible (when more than 40 per cent leaf and pod area were affected)

Range of important agronomic characters, oil content and fatty acid profiles in the oil are presented in Table 8.5. Both the mutant varieties were taller than the check which resulted more biomass yield that could be used as fuel for cooking purposes (Basak and Das, 1992). They also produced higher number of pods/plant. 1000-seed weight was higher in Agrani sharisha followed by Sonali. Maturity period of the mutant varieties and check was almost similar (85-95 days) except Tori-7 which matured about 7-10 days earlier than others. Improvement in oil content was found in Safal and Agrani sharisha (44.7 and 45.1 per cent oil) while the check, Sampad, Sonali and Tori-7 had 44.4, 44.6 and 41.8 per cent oil, respectively.

Moreover, both the mutant varieties had higher content of desirable oleic and linoleic acids and lower content of undesirable erucic acid in the oil compared to check. Undesirable linolenic acid content was more or less same in the mutant varieties and check. The above result agreed with the report of Ashri (1988) who found that characters like earliness, disease resistance, fatty acid composition in oil crops can be improved through induced mutation technique. Similar results were also reported by Kumar and Das (1977), Shaikh *et al.* (1982), Mahla *et al.* (1990), Kawai and Amano (1991), Robbelen (1991), Wang (1991) and many other researchers. Economic potential of Safal and Agrani sharisha was found to be higher in several studies (Qazi *et al.*, 1993; Costa *et al.*, 1995).

Table 8.5: Range of Important Agronomic Characters, Oil Content (%) and Fatty Acid Profile in Oil of Mutant Varieties Compared to the Existing Check Varieties

Variety	*Plant Height (cm)*	*Number of Pods/ Plant*	*Number of Peeds/ pod*	*1000-Seed Weight (g)*	*Days to Maturity*	**Oil content (%)*
Mutant Varieties						
Safal sharisha	150-180	68-120	18-26	2.7-3.0	88-95	44.7
Agrani sharisha	135-155	60-105	22-28	3.1-4.0	88-93	45.1
Check						
Sonali sharisha	115-136	50-89	18-26	3.1-3.8	86-95	44.6
Sampad	110-125	46-90	16-24	2.2-3.0	85-93	44.4
Tori-7	65-80	32-114	8-18	2.1-3.0	76-83	41.8

Variety/ check	*Palmitic Acid 16 : 0*	*Stearic Acid 18 : 0*	*Oleic Acid 18 : 1*	*Linoleic Acid 18 : 2*	*Linolenic Acid 18 : 3*	*Ecosenoic Acid 20: 1*	*Erucic Acid 22: 1*
Mutant Varieties							
Safal sharisha	2.6	1.1	15.6	17.0	8.8	8.7	46.7
Agrani sharisha	2.4	1.0	16.1	14.3	7.4	7.3	49.8
Check							
Sonali sharisha	2.3	0.7	15.4	12.4	8.9	5.3	48.1
Sampad	2.1	0.9	13.0	12.7	7.7	7.7	54.0
Tori-7	1.8	0.9	13.0	12.8	6.7	7.4	51.9

Source: Agronomic characters (Rahman *et al.*, 1992), Oil content and fatty acid profile (Annual Report, Oilseed Research Centre, BARI, 1992-93)

In 1991 the National Seed Board of Bangladesh, after a thorough field-evaluation, approved these two mutants as high yielding varieties tolerant to *Alternaria* blight for commercial cultivation throughout the country. The popular names of these mutants are Safal and Agrani sharisha. Farmers of various districts have started to grow the mutant varieties for getting higher economic benefit.

Improvement Made in a cv. Nap-3 of *B. napus*

All the M_1 plants survived at maturity were harvested and only the first formed five pods/plant were taken to raise the M_2 in plant progeny rows (Anon. 1991). A large number of mutants/variants were isolated from the M_2 segregating population in 1990-91 cropping season (Anon. 1992). However, only 27 true breeding mutants were selected in M_3 generation (Anon. 1993). Two mutants, MM19-33-90 and MM43-66-90 (later named as BINA sharisha-3 and BINA sharisha-4, respectively) were ultimately selected for further evaluation at different agro-ecological zones of the country during 1992-93 to 1996-97. The result is presented in Table 8.6. Mean of all the yield trials showed that BINA sharisha-4 produced significantly the highest seed yield (1.70 tons/ha) followed by BINA sharisha-3 (1.57 tons/ha). The check, BARI sharisha-7 and BARI sharisha-8 gave 1.43 and 1.40 tons seeds/ha, respectively. The result also showed that both the mutant varieties maintained all along (irrespective of year and location) superior performance in seed yield compared to check. The mutant MM36-41-90, inspite of its higher seed yield potential, was discarded for having lower oil content than the check.

In respect of maturity period, it was clearly observed that BINA sharisha-4 took 83 days, *i.e.*, 21 and 13-14 days early maturing compared to Nap-3 (Mother) and check varieties, respectively (Table 6). Similarly, BINA sharisha-3 took 87 days, *i.e.*, it matured 10-17 days earlier than Nap-3 and check.

The performance of two mutant varieties and check was also tested under recommended management practice (high input conditions) and farmers' management practice (low input conditions) in the farmers' field at two agro-ecological zones during 1995-96 and 1996-97 seasons. BINA sharisha-3 and BINA sharisha-4 have proved their superior performance in both the management practices (Table 8.7). The recommended cultural practice gave significantly higher seed yield (1.55 tons/ha) than the farmers' practice (1.45 tons/ha).

Table 8.6: Seed Yield of Two Mutant Varieties of *B. napus* Compared to Check During 1993-94 to 1996-97 in Eight Different Agro-Ecological Zones of Bangladesh

Variety/Mutant/ Check	*Seed Yield (tons/ha)*						
	**PYT at 2 AEZ in 1992-93*	*SYT at 3 AEZ in 1993-94*	*AYT at 3 AEZ in 1994-95*	*RYT at 5 AEZ in 1995-96*	*ZYT at 6 AEZ in 1996-97*	*Mean*	*Days to Maturity*
Mutant Varieties							
BINA Sharisha-3	1.46b	1.59b	1.51b	1.47b	1.82a	1.57b	87
BINA Sharisha-4	1.57a	1.74a	1.74a	1.66a	1.81a	1.70a	83
Elite Mutants and Check							
MM36-41-90	1.36c	1.61b	1.53b	1.34c	—	1.46c	95
MM15-94-90	1.22d	1.35c	1.29c	—	—	1.29d	97
Nap-3 (Mother)	0.98e	1.04d	1.34c	1.26c	—	1.16e	104
BARI Sharisha-7	—	—	—	1.26c	1.59b	1.43c	97
BARI Sharisha-8	—	—	—	1.17d	1.62b	1.40c	96

Values having common letter (s) in each column do not differ significantly at 5 per cent level.

*PYT = Preliminary Yield Trial, SYT = Secondary Yield Trial, AYT = Advanced Yield Trial, RYT = Regional Yield Trial, ZYT = Zonal Yield Trial.

Table 8.8 shows that oil content in the mutant varieties was higher (44 per cent) than Nap-3 (42 per cent). Desirable changes in fatty acid profiles were found in the mutant varieties compared to Nap-3 (Mother). Substantial increase in desirable oleic acid and decrease in undesirable erucic acid were found in BINA sharisha-3 and BINA sharisha-4 which had 27.4 and 36.6 per cent oleic acid, and 26.4 and 20.0 per cent erucic acid, respectively whilst Nap-3 (Mother) had 21.8 per cent oleic acid and 34.9 per cent erucic acid as reported elsewhere (Das *et al.*, 1995). Improvement obtained in seed yield, earliness, oil content and fatty acid quality in the present investigation is in agreement with the achievements of Shaikh *et al.* (1982), Hakim *et al.* (1988), Robbelen (1991) and Zakri (1991).

Table 8.7: Seed Yield of Two Mutant Varieties Compared to Check During 1995-96 and 1996-97 in the Farmers' Field at Two Agro-Ecological Zones (AEZ) of Bangladesh

Year, AEZ and Variety	**RM*	*% Increased Over*		***FM*	*% Increased Over*	
		BARI Sharisha-7	*BARI Sharisha-8*		*BARI Sharisha-7*	*BARI Sharisha-8*
		1995-96 AEZ, Dinajpur				
Mutant Varieties						
BINA Sharisha-3	1.67a	20.1	24.6	1.55a	17.4	21.1
BINA Sharisha-4	1.46b	5.0	9.0	1.34b	1.5	4.7
Check						
BARI Sharisha-7	1.39b		—	1.32b	—	—
BARI Sharisha-8	1.34b		—	1.28c	—	—
		1996-97 AEZ, Jessore				
Mutant Varieties						
BINA Sharisha-3	1.65b	13.8	3.1	1.54b	10.8	4.8
BINA Sharisha-4	1.79a	23.4	11.9	1.66a	19.4	12.8
Check						
BARI Sharisha-7	1.45c	—	—	1.39bc	—	—
BARI Sharisha-8	1.60b	—	—	1.47b	—	—
Mean of 2 AEZs						
BINA Sharisha-3	1.66a	17.0	13.9	1.55a	14.1	13.0
BINA Sharisha-4	1.63a	14.2	10.5	1.50a	10.5	8.8
BARI Sharisha-7	1.42b	—	—	1.36b	—	—
BARI Sharisha-8	1.47b	—	—	1.38b	—	—
Mean of 2 Managements	1.55a	—	—	1.45b	—	—

Values having common letter (s) in each column and year do not differ siginificantly at 5 per cent level.

*RM = Recommended cultural management, **FM = Farmers' cultural management.

In 1997 the National Seed Board of Bangladesh has approved (registered) the two mutants as early maturing and high yielding varieties for commercial cultivation in the farmers field all over the country. The popular names of these mutant varieties are BINA sharisha-3 and BINA sharisha-4.

Table 8.8: Fatty Acid Profiles of BINA Sharisha-3, BINA Sharisha-4 and Nap-3 (Mother)

Mutant/ Mother	*Oil Content (%)*	*Palmitic Acid 16 : 1*	*Stearic Acid 18 : 0*	*Oleic Acid 18 : 1*	*Linoleic Acid 18 : 2*	*Linolenic Acid 18 : 3*	*Erucic Acid 22 : 1*
BINA Sharisha-3	44.0	5.1	1.6	27.4	14.7	21.3	26.4
BINA Sharisha-4	44.0	5.2	1.9	36.6	15.6	21.0	20.0
Nap-3 (Mother)	42.0	5.0	1.6	21.8	14.0	21.8	34.9

Source: Modified after Das *et al.* (1995).

In conclusion, it can be said that induced mutation technique (gamma rays) effectively created genetic variability in two cultivars (YS-52 and Nap-3) of *Brassica* oil crops which made a good scope for selection of desirable mutants, and ultimately two yellow seeded varieties of *B. campestris* and two brown seeded varieties of *B. napus* were developed and released for commercial cultivation.

REFERENCES

Anonymous. 1977. Manual on Mutation Breeding. 2nd Edn. Joint Division of FAO/IAEA. International Atomic Energy Agency, Vienna. Technical Report Series No. 119. pp. 51-78.

Anonymous. 1991. Annual Report, 1989-90. Bangladesh Institute of Nuclear Agriculture, Mymensingh. pp. 13 & 18.

Anonymous. 1992. Annual Report, 1990-91. Bangladesh Institute of Nuclear Agriculture, Mymensingh. p. 13.

Anonymous. 1993. Annual Report, 1991-92. Bangladesh Institute of Nuclear Agriculture, Mymensingh. pp. 15-18.

Ashri, A. 1988. Mutagenic studies and breeding of peanut (*Arachis hypogaea*) and sesame (*Sesamum indicum*). Proc. Workshop on the Improvement of grain legume production using induced mutations. Held at Pullman, Washington, USA from July 1-5, 1986. FAO/IAEA, Vienna. pp. 509-511.

Basak, N.C. and Das, M. L. 1992. Performance of some mutant strains of mustard under two managements. Bangladesh J. Nuclear Agric. 7&8: 85-89

Costa, D.J., Qazi, A.K. and Das, M. L. 1995. Performance of different mutant lines of mustard in coastal areas under late sown conditions. Bangladesh J. Training and Development 8 (1 & 2): 69-72.

Das, M. L. and Rahman, A. 1994. Evaluation of mutants of rapeseed for earliness, seed yield and response to changing environments in M_5 generation. Bangladesh J. Nuclear Agric. 10: 75-80.

Das, M. L., Rahman, A., Malek, M.A. and Pathan, A.J. 1995. Stability and fatty acid analyses for selecting promising mutants of rapeseed. Bangladesh J. Nuclear Agric. 11: 1-8.

Hakim, L., Azam, M.A., Miah, A.J. and Mansur, M.A. 1988. Promising rice mutants. J. Nuclear Agric. Biol. 17: 125-127.

Kawai, T. and Amano, E. 1991. Mutation breeding in Japan. In: Mutation Breeding for Crop Improvement. Vol. 1. Proc. Symp. FAO/IAEA, Vienna, 18-22 June, 1990. pp. 47-66.

Kumar, P.R. and Das, K. 1977. Induced quantitative variation in self compatible and self incompatible forms of *Brassica*. Indian J. Genet. Pl. Breed. 37 (1): 5-11.

Mahla, S.V.S., Mor, B.R. and Yadav, J.S. 1990. Induced genetic variability for oil content in mustard (*Brassica juncea* L. Czern & Coss). Oil Crops Newsletter, IDRC, Ottawa, Canada, No. 7. pp. 13-15.

Micke, A., Donini, B. and Maluszynski, M. 1987. Induced mutations for crop improvement-a review. Trop. Agric. 64: 259-278.

Novak, F.J. and Brunner, H. 1992. Plant breeding: Induced technology for crop improvement. IAEA Bull. 4. pp. 25-33.

Panda, B.S. 1981. Success in mutation breeding of sesame. Mutation Breeding Newsletter, IAEA, Vienna. Issue No. 18: pp. 10-11.

Qazi, A.K., Costa, D.J., Das, M.L. and Akbar, M.A. 1993. Evaluation of different varieties/advanced lines of rapeseed and mustard under late sowing conditions in Patuakhali area. Bangladesh J. Agril. Sci. 20 (2): 299-302.

Rahman, A., Das, M.L. and Howlider, M.A.R. 1988. Induction of genetic variability in mustard for selection against *Alternaria* blight. Bangladesh J. Nuclear Agric. 4: 30-36.

Rahman, A., Das, M. L. and Pathan A.J. 1992. New high yielding mutant varieties of mustard (*Brassica campestris* L. var. Yellow Sarson). J. Nuclear Agric. Biol. 21 (4): 281-285.

Robbelen, G. 1991. Mutation breeding for quality improvement: A case study for oilseed crops. In: Mutation Breeding for Crop Improvement. Vol. 2. Proc. Symp. FAO/IAEA, Vienna, 18-22 June, 1990. pp. 3-30.

Shaikh, M.A.Q., Ahmed Z.U., Majid, M.A. and Wadud, M.A. 1982. A high yielding and high protein mutant of chickpea (*Cicer arietinum* L). derived through mutation breeding. Environment and Exptl. Bot. 22: 483-489.

Steel, R.G.D. and Torrie, J. H. 1960. *Principles and Procedures of Statistics.* McGraw-Hill Inc. International Student Edn., Tokyo. Pp. 195-208.

Tsunoda, S. 1980. Eco-physiology of wild and cultivated forms of *Brassica* and allied genera. In: *Brassica* crops and wild allies. Japan Sci. Soc. Press, Tokyo.: pp. 109-119.

Wang, L.Q. 1991. Induced mutation for crop improvement in China-A review. In: Mutation Breeding for Crop Improvement. Vol. 1. Proc. Symp. FAO/IAEA, Vienna, 18-22 June, 1990. pp. 9-32.

Zakri, A.H. 1991. Breeding high yielding soybean varieties using induced mutation. In: Mutation Breeding for Crop Improvement. Vol. 2. Proc. Symp. FAO/IAEA, Vienna, 18-22 June, 1990. pp. 163-170.

Chapter 9

Effects of Gamma Radiation on *Jatropha curcas*: A Promosing Crop for New Source of Fuel

S.K. Datta & R.K. Pandey

Botanic Garden and Floriculture,
National Botanical Research Institute, Lucknow, (U.P.), India

ABSTRACT

Jatropha curcas L. oil has been identified as an efficient substitute fuel for diesel engine. To increase its oil content, seeds were treated with different doses of gamma rays. Delay in germination and reduction in plant height, branch and leaf number and different types of morphological abnormalities were observed after irradiation. Wide range of variability could be induced in plant height, seed production and oil content. Induced mutagenesis was found to be promising tool for developing both tall and dwarf, high branch, high fruit and oil yielding and high biomass yielding mutant.

INTRODUCTION

Since the oil crisis in 1974, most oil importing countries have been highly motivated to develop alternative sources of energy to meet their natural energy needs. A number of options for alternative liquid fuel production have been considered in many countries. To overcome the energy problems a great deal of research on solar energy, wind energy, hydropower, biomass energy and biogas has been carried out by government organizations and educational institutes in different countries. The oil seed option could complement alcohol fuel for petrol engines. In view of the uncertainties in the availability and price of liquid fuels from petroleum it would appear prudent to carry out research and development of seed oil as alternative fuels for diesel engines in order that their production and use could be rapidly implemented, should the need arise. For this reason, use of agriculture products to provide renewable sources of energy has appeal. Subsequently an assessment was made for the potential to use vegetable oils as substitute for diesel fuels. A number of options for alternative liquid fuel production have been considered in many countries. Some are based on processing of fossil fuels such as coal, shale and natural gas reserves while others are based on biomass, a renewable source of fuel. The potential production of ethanol and methanol from crops and crop residues have been evaluated (Stewart *et al.*, 1981). The oil seed option could complement alcohol fuel alternatives, in that the seed oil are more suited for diesel than for petrol engines. Tests were made with diesel engines using *Jatropha curcas* L. Oil as part of a research project carried out with the co-operation of Yanmar (Thailand) Co. Ltd. and very satisfactory results were obtained. The engine performance and fuel consumption compared favourably with running the engines on normal diesel engine oil (Bhasabutra and Sutiponpeibun 1982a, b, Jaray 1984, Martin and Mayeux 1984, Munch and Keifer 1989, Sharma 1985, Takeda 1982a, b, Takeda and Minoru 1981). Utilization of *J. Curcas* oil as a new source of oil for diesel engine is in the very early stage and there are only a few published literature available to answer the many specific questions about its production and commercialization. The present paper reports the results of effects of gamma radiations possibilities for and induction of desirable variations through induced mutagenesis.

Jatropha Curcas (plant family Ehphorbiaceae) oil has been reported as an efficient substitute fuel for diesel engines. The engines performance and fuel consumption compared favourably with running the engines on normal diesel engine oil. During exhaust gas test it has been observed that engine when run with curcas oil, the value of carbon monoxide and the smoke was lower than the accepted value as per the standard specification of the environment board. There is no emission of sulphur dioxide in curcas oil exhaust gas as compared to 125 ppm in diesel engine exhaust gas. The parameters of *Jatropha* biodiesel as reported by Foild and Eder (1997) and Foidl *et al.* (1996) are similar to the biodiesel from rapeseed oil, currently widely used in any automobile brand which has been approved for its use by the manufactures without any modification to the engine or accessories. Biodiesel has similar or better fuel consumption, horse power and torque and haulage rates as conventional diesel.

J. Curcas a tropical species has naturalised in India in several areas and is only cultivated as a hedge around cutlivated field in a semiwild condition. It was introduced in India by Portuguese as an oil yielding plant. It is commonly known as ' dravanti', jangliarandi', 'ratanjota', 'baghbherenda', jamalgota', 'kattamanakku', 'kacha' etc. in different states of India. It is known as physic nut and purging nut in English. Different names are given in different countries like 'sabudam' in Thailand, 'thinbaukyeksu' and 'thinbaukyekku' in Burma, ' tuba' in Phillipines, 'ma fong chou' in China, 'kadam' in Nepal, 'pinheiro de purga' in Brazil, 'ricin d' Amerique' in French and ' roppendaru' in Sri Lanka. It is one of the promising drought tolerent perennial crop. It is adaptable to various kinds of soil conditions. It grows well on moderately sodic and saline degraded and aroded soils. An average seed production of about 5 tonne per hectare can be expected under optimum conditions. A yield of 0.75 to 2 tones of biodiesel can be expected per hectare per year from the fifth year onwards. The additional advantage of *J., Curcas* is that its oil is nonedible and capable of growing on waste lands so that their production does not exert any pressure on agricultural land. It can easily be grown both by seed and by stem cuttings. It grows rapidly, is hardly to dry weather conditions and is not browsed by goats or cattle. Its seed contains 40-50 per cent semi-drying oil of pale yellow colour with acrid taste and 70 per cent moisture. The oil contains about 21 per cent saturated fatty acids and 70 per cent unsaturated fatty acids; palmitic, stearic, oleic and linoleic acids being the major fatty acids. The oil cake also contains high organic substances and it can be utilized as fertilizer. The oil can be used as an illuminant, it burns without emitting smoke. Rural people can directly use the oil for lighting lamps. It can be used also as a lubricant and for making soaps and candles. The latex can be used for making resin and dye.

No regular seed collection of this plant has been done for any end uses in our country. The seed oil productivity is very low at present since it has never been improved for oil crops and the basic knowledge for domestication is quite limited. Systematic efforts have not been made for better cultivation and for increase in production by using present day knowledge of plant breeding, increased agronomic experience and induced mutagenesis.

The characterization of 'curcas oil' and utility research to decide on its alternative industrial application, and the overall plant productivity and yield figures to develop economically feasible husbandary of Jatropha has not received due attention. No recent data is available on systematic approach for biological assessment for cultivation, growth pattern, branching habit, spacing, pruning operation, total yield, sensitivity to matagenesis for development of promising varieties. National Botanical Research Institute (NBRI), Lucknow has already made basic studies for biological assessment for cultivation, growth pattern, branching habit, spacing, pruning operation and total yield. NBRI has also tested sensitivity of Jatropha seeds to different mutagens (gamma radiation, colchicine) for developing high yielding strains through genetic manipulations. Government is now talking proper interest for developing such alternative fuel source.

Materials and Methods

Dry seeds of *Jatropha curcas* Linn, (Euphorbiaceae) were exposed to 6, 12 and 18 Krad gamma rays (Cobalt–60, radiation source) and sown in randomized block designed beds. Equal number of unirradiated seeds were also sown which served as control. Data were recorded on germination, growth and other morphological and seed characters.

Results and Discussion

Seed germination was reduced after treatment with 18 Krad gamma rays. Germination percentage was more due to stimulation after treatment with 6 and 12 Krad over the control. Delay in seed germination was recorded after exposure to 12 Krad and 18 Krad and the delay in both the cases was significant (P<0.01). Seedling height was recorded in control and all the irradiated populations on 1st, 3rd and 4th month after sowing. Reduction in seedling height was recorded in all the treatment. Significant (P<0.01 to P<0.01) reduction in average plant height was recorded at 3rd and 4th month in both 12 and 18 Krad treated populations (Table 9.1). Wide range of variability in plant height was observed in the irradiated populations which was clear when height of individual plant was analysed at different growth period. The growth of few treated plants was stunted and few others were vigorous than the control.

Table 9.1: Effects of Gamma Rays on *Jatropha curcas*

	0	*Gramma Rays (Krad)*		
	(Control)	*6*	*12*	*18*
Germination (%)	56.25	68.75	62.50	50.00
Germination time (days)	5.33 +−0.44	5.27 +−0.49	7.00** +−0.26	7.25** +−0.31
Ist month height (cm)	13.55 +−1.64	13.36 +−0.77	11.80 +−0.68	10.62 +−0.82
3rd month height (cm)	69.11 +−4.64	60.91 +−7.30	50.50** +−4.52	30.30** +−2.022
4th month leaf No.	102.67 +−6.16	101.90 +−8.01	85.40 +−5.45	49.94 +−4.49
2nd month leaf	66.44 +−6.69	81.45 +−9.77	36.60** +−4.83	26.00*** +−3.72
2nd month leaf abnormalities	11.37 +−1.30	21.43*** +−1.37	22.40*** +−2.18	27.88*** +−3.11
4th month leaf No.	150.78 +−6.93	149.82 +−22.12	85.70*** +−12.26	51.25*** +−5.05
4th month leaf abnormalities (%)	20.41 +−1.09	15.65 +−0.89	30.34 +−1.57	49.02 +−2.47
Primary Branch No. (4 month)	3.22 +−0.22	2.18 +−0.32	1.40*** +−0.30	1.25*** +−0.16
Secondary branch No. (4 month)	6.75	2.00	2.83	3.16
Total Branch (21 month)	10.12 +−1.53	9.36 +−1.80	11.50 +−0.80	7.25 +−0.94
% plants with secondary branch	44.44	72.73	60.00	62.50

Contd...

Table 9.1–Contd...

	0	Gramma Rays (Krad)		
	(Control)	6	12	18
Leaf width (cm)	15.46 + – 0.42	15.02 + – 0.40	14.36 + – 0.38	12.66** + – 0.30
Leaf width (cm)	17.90 + – 0.32	17.76 + – 0.39	17.62 + – 0.36	16.12** + – 0.37
Petiole length (cm)	22.80 + – 0.33	21.05 + – 0.55	21.52 + – 0.53	17.34** + – 0.42
Stem diameter	26.62 + – 1.01	26.13 + – 1.50	24.30 + – 0.92	22.44 + – 0.79
Stem diameter (cm) 21 months	30.62 + – 1.18	31.00 + – 1.76	28.90 + – 0.85	27.00 + – 1.10
No. of seeds Per plant	112.00 + – 30.96	63.50 + – 21.93	21.16+ + – 12.15	12.33** + – 4.09
Seed length (cm)	1.728 1.018	1.712 + – 0.017	1.658 + – 0.019	1.670 + – 0.015
Seed diameter	1.075 + – 0.010	1.10 + – 0.017	10.54 + – 0.019	1.076 + – 0.015
Single seed Wt. (g)	0.533 + – 0.040	0.560 + – 0.026	0.467 + – 0.043	0.458 + – 0.031
10 seed Wt. (g)	5.281 + – 0.132	5.840*** + – 0.027	5.123 + – 0.223	4.750*** + – 0.312
Kernel Wt. (g)	0.403	0.393	0.300	0.312
Oil (%)	40.70 + – 1.33	39.53 + – 1.26	36.74 + – 1.94	31.20*** + – 0.01

= P < 0.05, + = P< 0.02; ** = P < 0.01; *** = P< 0.001

The average leaf number per plant was significantly (P<0.01 to P< 0.001) reduced after treatment with 12 and 18 Krad at 2nd month and increase in leaf number was recorded after treatment with 6 Krad. At 4th month age significant (P<0.001) reduction in leaf number was recorded in 12 and 18 Krad treated populations. Different types of abnormalities in leaves were observed in the treated populations. The leaf abnormalities included changes in shape, size, margin, apex, fission and fusion of leaves. The leaf abnormalities increased significantly (P<0.0001) after irradiation and with increase in exposure at 2nd month over the control. At 4th month the leaf abnormalities in 12 and 18 Krad were significantly (P<0.001) more over the control. Wide range of variability in leaf number was also recorded in individual plant analysis in the treated populations. Branch number was reduced after irradiation and with increase in exposure. The reduction was significant (P<0.001) after exposure to 12 and 18 Krad. Leaf size (Length, width and petiole) was decreased after irradiation and with increase in exposure. Leaf length, width and petiole was significantly (P<0.1) to P<0.001) reduced after exposure to 18 Krad (Table 9.1). There was no seed formation in the first year. All the plants were carefully maintained for observation in subsequent year. In the 2nd 3rd year the main emphasis was given on fruits and growth of the plants.

In the 2nd year the average height of plants of treated populations was less than the control. The height of 18 Krad treated populations remained significantly (P<0.001) reduced. In the 3rd year the same 18 Krad treated populations remained significantly (P<0.01) reduced. In 3rd year few plants of 6 Krad treatment were taller than the control and therefore the average height was more in 6 Krad treated population over the control. Total branch at 12 Krad and plant height and stem diameter at 6 krad on 21 month was more over the control. Stem diameter was less in the treated populations at the age of 14 and 21 month except at 21 month the stem diameter was more in 6 krad treated populations. Although the average value of these characters were not significantly more over the control but individual plant analysis showed significant increase of these characters in some treated plants over the control.

In the 3rd year there were seed formation and the average seed production per plant was reduced in the treated populations. The seed size in higher doses was reduced but the diameter of seed in 6 Krad treated population was more in comparison to control. Individual seed weight was less in 12 and 18 Krad populations but in 6 Krad populations was significantly (P<0.001) more over the control. The percentage of oil and kernel weight decreased after treatment with 12 and 18 Krad. In the 6 krad treated populations oil content per seed and kernel weight per seed analysis showed wide range of variability than in the control. As the oil was present mainly in Kernel, an attempt was made to study the relationship of seed and kernel weight to the oil content. It was observed that the more was the seed and kernel weight the more was the oil content.

A wide range of variability could be induced in the present experiment by gamma irradiation. Selections for variants with high oil yield and increased branch number, seed weight and seed yield, based on improved performances of the treated plants over the best in the control, were made in the second generations after treatment and studied. Some of the selections showed true breeding nature of their desired character even after one year of the growth. All the selected lines have been carefully maintained to study their true breeding nature and performance in further generations. Desirable selected lines are beneficial for direct use in the industry and in agriculture for future breeding programme.

The effects of gamma rays on *J. curcas* are in conformity with the results obtained in a large number of plants by many investigators. The cause of induced variation in the first generation after mutagen treatment has been very critically analysed by different workers and reported from time to (Datta 1988; Datta and Basu 1988; Gunckel and Sparrow 1961; Konzak 1957; Sparrow et at. 1952). X-ray has been successfully used for inducing high oil and punicic acid content mutant in Trichosanthes anguina (Datta 1989). From the analysis of data it is found that there are possibilities to isolate different types of mutants from the present experiment. Both tall and dwarf, high branch, high fruit and oil yielding and high biomass yielding mutant can be established.

Acknowledgement

Thanks are due to the Director, National Botanical Research Institute, Lucknow, India for providing the facilities. The work was carried out under a project supported by Department of Non-Conventional Energy Sources, Govt. of India, New Delhi which is thankfully acknowledged.

REFERENCES

Bhasabutra, R. and Sutiponpelbun, S.1982a, Jatropha curcas oil as a substitute for diesel engine oil. Renewable Energy Rev. J. 4: 56–70.

Bhasabutra, R. and Sutiponeibun, S.1982b. The study of Jatrophas curcas oil as a substitute of diesel engine oil. Thai Verglon, Department of Agriculture, Ministry of Agricultural and Co-operatives, Bangkok. 42p.

Datta, S. K.1988. Induced cytomorphological changes in X, and C generations of three members of Cucurbita-ceae, J Indian bot. Soc. 67 271-274.

Datta, S. K. 1989. Induced high punicic acid yielding mutans on Trichosanthes anguina. J. Indian bot.Soc. 68 44-46.

Datta, S. K. and R. K. Basu 1988. Abnormal plant growth in M and C generations of tree members of Cucurbitaceae. J Indian bot. Soc. 67 27-274.

Foidl N, Foidl G., Sanchez M, Mittelbach M, Hackel S. 1996. Jatrophas curcas L. as a source for the production of biofuel in Nicaragua. Bioresource Technol. 58, 77-82.

Foild N, Eder P. 1997. Agro–industrial exploitation of J. curcas. In, Biofuels and Industrial Products from Jatropha curcas, Gubitz GM, Mittelbach M., Trabi. M. (Eds), Dbv–Verlag, Graz, Austria.

Gunckel, J. E. and Sparrow, A.K. 1961. Ionixing radiation; Biochemical physiological and morphological aspects of their effects on plant, Encyclopedia of Plant physiology 16 555-617.

Jaray, Sadakorn.1984. Potential of physic nut (Jatrophas curcas Linn). as an energy source in Thailand. Agric. Res. OK. (Thailand), 2 (1): 67-72.

Konzak, C. E. 1957. Genetic effects of radiations on higher plants, Q. Rev. Biol. 32 27-45.

Martin, G. and Mayeux, A. 1984. Reflection on oil crops as source of energy. II.The physic nut (Jatropha curcas L).. A possible source of fuel. Oleagineux, 39 (5): 283-87.

Munch, E. and Keiefer, J. 1989. (Physic nut: a multipurpose plant as a future source of motor fuel) Die purgiernuss (J. curcas): Mehrzweckpflanze als kraftsoffquelle der zukunft". Zusammenarbeit. 209: 32.

Sharma, S.C. 1985. Jatrophas curcas–a possible substitute for diesel oil. PTI Sci. Ser., 18.

Sparrow, A. H., M. J. Moses and Dubow, R., 1952. Relationships between ionizing radiations, chromosome breakage and certain other nuclear disturbances, Exptl. Cell Res. Suppl. 2 245–267.

Stewert, G.A., Rawlins, W.H.M., Quick, G. R., Begg, J.E. and Peacock, W.J.1981. Oil seeds as a renewable source of diesel fuel. Search, 12; 107-15.

Takeda, Y. 1982a. An on –going study of Jatropha curcas oil as a substitute for diesel engine fuel in Thailand. Selected Solar Energy Tech., Oct., 4 (9): 21-24.

Takeda, Y. and Minoru, O. 1981. Interim Report on the study of Jatropha curcas oil as a substitute for diesel engine oil. Industrial Finance Corporation, Thailand, Bangkok, Oct., 14.

Takeda, Y.1982b. Development study on Jatropha curcas (Sabudam) oil as a substitute for diesel engine oil in Thailand. J. agric. Assoc., China, 120: 1-8.

Chapter 10

Role of Mutation Induction for Wheat (*T. aestivum* L.) Improvement

Wange Lin-quing & LI Gui-ying

Institute for Application of Atomic Energy (IAAE), Chinese Academy of Agricultural Sciences (CAAS), P.O. Box 5109, Beijing – 100 094, China

Foreword

Wheat (*Triticum* L). is the most important food crop in the world, and second only to rice in China. According to statistics of FAO (Production Yearbook, 1990, Jin 1996), the annual area and output, and average yield per hectare of wheat in the world shown in Table 10.1.

Wheat belongs to *Triticum, Gramineae* as well as self–pollination crop. Based on the growth habit, genome and chromosome number, they could be classified into spring wheat, winter wheat, and semi-winter wheat. Regarding species of the *Triticum*, there are 22 species listed in (Table 10.2).

Table 10.1: Annual Cultivation Area and Production in Main Countries and China in 1989

Countries	*Annual Cultivation Area (kha)*	*Average Yield (Kg/ha)*	*Annual Total Output (kt)*
Main countries of Wheat production	221846	2337	518495
China	29141	3025	88102
%	13.16	129.44	16.99

Note: Main countries of wheat production are former USSR, China, India, Australia, Canada and France.

Table 10.2: Classification of *Triticum* (Dong Yushen, 1996)

Line	*Genome*	*Species*	*Form*
Einkorn	A	*T.uraatu thum*	Wild
		T.boeoticum Boiss	Wild
		T.monococcum L.	
Emmer	ABD	*T.dicccoides Koer*	Wild
		T.dicoccum	
		T.paleocoolchicum Men	
		T.isphanicum Nevski	
		T.turgidum Desf	**
		T.turanicum Jakubz	*
		T.polonicum L.	
		T.aethiopicum Jakubz	
Dinkel		*T.spelta L.*	
		T.macha Dek. Et Men	
		T.vavilovi Jakubz	
		T.compactum Host	*
		T.sphaerococcum Porc	
		T.aestivum L	**
Timopheevil	AG	*T.araraticum Jakubz*	Wild
		T.timopheevii	*
Zhukovskyi	AAG	*T. Zhukovskyi Men at Er.*	

Note: **, very important species fro production. *, Some useful species for breeding

Among these species, common wheat (*T. aestivum* L). is widely studied and used in breeding program and agricultural production, as well as in mutation breeding. It will be discussed in this chapter.

To develop and adopt good cultivars of wheat is a very economic and effective way for raising wheat production. The approaches of wheat improvement, such as introduction, pedigree selection, crossbreeding, biotechnique breeding, wide-hybridization and mutation breeding etc., have been widely used in breeding program.

Induced mutation generates genetic variability, promotes genetic recombination. Mutation breeding of wheat has been practiced in the world since 1950s. Use of certain special problems of breeding. It has become an effective way for supplementing existing gremplasms and improving cultivars, and recognized as one of the driving forces of evolution (Micke *et al.*, 1987). Mutation induction is also an important supplement to conventional breeding and would be difficult to substitute them with other methods. Many mutant varieties and hundreds of various valuable strains have been developed and released for production and used in cross breeding, thus making a substantial contribution.

According to the data of "Mutation Breeding Newsletter" of IAEA, after the first mutant variety of wheat, EIS with dwarf stem and resistance to lodging and good bake quality, derived from X-ray irradiation at Weihenstphan Institute of Agriculture and Cultivation of Plants, Germany in 1960. Famous mutant varieties of durum wheat, such as, Castelporziano, castelfusano, Creso, Mida, Tito, Giano and Augusto etc., were bred at the Centro Ricerche Energia Casaccia, Italy, during the 1970s and 1980s from direct use and indirect use with mutants induced (Savascia-Mungnozza *et al.*, 1991).

Wheat improvement by induced mutation has been practiced in China since 1957 and substantial achievements have been made during the past 40 years (Wang 1991, 1995,1996)· According to incomplete statistics in 1998, more than 111 mutant cultivars, about ¼ of the total mutant cultivars of plant crops, 10 per cent of the total wheat cultivars, which derived from other various breeding approaches, have been developed and released (or approved) for cultivation and commercial production. They covered about three million hectares, thus making a substantial contribution toward the national economy. Most mutant cultivars were directly utilized mutants, accounting for 82.0 per cent , 18.0 per cent were resulted from crossing with induced mutants (Table 10.3 and Fig. 10.1)

Table 10.3: Number of Released Mutant Cultivars of Wheat in China During Past Years (Wang L.Q., 1998)

Year	*Mutants Used*				*Total*	
	Directly		*Cross with them*			
	No.	*%*	*No.*	*%*	*No.*	*%*
1966-1970	14	12.6	14	12.6		
1971-1975	9	8.1	4	3.6	13	11.7
1976-1980	18	16.2	2	1.8	20	18.0
1981-1985	13	11.7	4	3.6	17	15.3
1986-1990	17	15.3	3	2.7	20	18.0
1991-1995	12	10.8	4	3.6	16	14.4
1996-now	8	7.2	3	2.7	11	14.4
Total	**11**	**82.8**	**20**	**18.0**	**111**	**100.00**

More than 30 mutant cultivars with outstanding characteristics, which were cultivated over 70kha/year (Table 10.4)

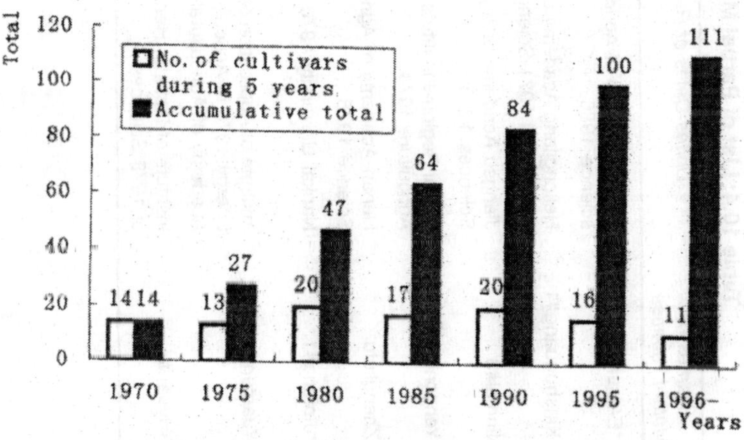

Fig. 10.1: Number, Date (Year) of Mutant Cultivars of Wheat (Wang. L.Q, 1998)

Table 10.4: List of Partial Mutant Cultivars of Wheat Planting Over 70 Kha per Year

Names of Mutant Varieties	*Unites Dates of Release*	*Mutagens and doses [Initial Material]*	*Main Characters Improved*
**E-mai#6	Hubei academy of Agricultural science 1986	γ 300 Gy [Nanda2419]	Rusts resistance, wide adaptation, high and stable yield.
*Xinshuguang#1	Heilongjiang Academy of Agricultural 1971 Sciences	γ 80Gy [AboM4/Oro] F4	Resistance to lodging and disease tolerance to wet and low temperature.
Ninmai#3	Jiangsu Academy of Agricultural Sciences 1973	γ 250Gy [st 1472/506]	Dwarf, lodging resistance, rusts resistance, high yield.
*Yennong6.85	Yantai Regional Institute of Agriculture 1974	[Youbo/fuxi#4]	Cold tolerance, leaf rust resistance
*Zhengliufu	Henan Academy of Agricultural Science 1976	γ 300Gy [Zhengzhou#6]	Drought and cold tolerance, shorter stem, leaf rust resistance.
**Jinfang#1	Nankai University 1976	γ 250G [Shijiazhuang63]	5-7 days earlier 20-35cm shorter than the original material, high yield.
*Yuanfeng#4	Institute of Application of Atomic Energy 1976 In Agriculture, Shandong Academy of Agricultural Sciences	γ 300Gy[Taishan#4]	Dwarf, lodging resistance, high yield.
*Yuyuan#1	Institute of Isotope, Henan Academy of 1979 Sciences	γ 350G [st2422/464/ Neixiang#5]	Early, dwarf, tolerance to stresses, good adaptation, high yield

Contd...

Table 10.4–Contd...

Names of Mutant Varieties	*Unites Dates of Release*	*Mutagens and doses [Initial Material]*	*Main Characters Improved*
***Shannounfu#63	Shandong Agricultural University 1980	γ 300Gy [Youbo/Oro] F4	Early, wide adaptation and good yield
*Chuanfu#1	Insitute of Bio-Nucl. Technique, Sichuan 1982 Academy of Agricultural Sciences	B ray 10 μ c/grain [Chuanyu#5]	Early, strip and rust resistance, drought and cold tolerance
*Xinchun#2	Insitute for Application of Atomic Energy, 1984 Xinjiang Academy of Agricultural Sciences	γ 80 Gy [Sailosi/Qichun#4]	Early, good quality, high yield
**Yuandong#3	Instiute for Application of Atomic Energy, 1989 Chinese Academy of Agricultural Sciences	γ 250G [12040/Auvror] F_3	Complex resistance of rusts, powdery mildew, drought, salt-Kali tolerance, wide adaptation
**Xinchun#6	Instiute for Application of Atomic Enegy, 1993 Xinjiang Academy of Agricultural Sciences	[Zhong/Xinchun#2]	High and stable yield, good quality and adaptation

Note: *, Cultivated area more than 70 Kha. **, Cultivated area more than 200 Kha. ***, Cultivated area more than 700 Kha.

A lot of various desirable mutants have been obtained and used for wheat improvement. Depending on the needs of production and breeding of wheat, several hundreds variously desirable mutant strains, including earliness, dwarfism, disease resistance, tolerance to environmental stress, and good quality etc., have been obtained, collected, preserved in mutant gremplasms bank and used in breeding program as well as in genetic studies. Some outstanding mutants are listed as follows (Table 10.5, Wang 1995).

Table 10.5: List of Various Mutant Strains with Special Features

Names of Mutant Strains	*Special Features*	
Longfumai#1	Earliness	70-75 days duration
Lumai#20		10-15 days earlier than the original
Fuai#1	Dwarfism	60 cm height, dwarfed by 40 per cent , controlled by 2 a dominant gene of dwarf
Yuanding#96		75 cm height, dwarfed by 20 per cent , controlled by 2 allelic genes (2D,6D)
Yuanding#3	Resistance	Rusts, powdery mildew resistance; tolerance to alkali saline and drought etc.
93K809		Scab resistance
Gannongfu	Quality	Protein content 21.0 per cent
85EA and 89AR	Cytoplasmic male sterile	Fertility easier to recover and restore resource widely
Fu#66 etc.	Combining ability	Very good combining ability for crossbreeding

Although the desirable traits mentioned above were available in wheat gremplasms collection, breeders prefer to induce and use them for wheat breeding. Use of mutants in crossbreeding has proved to be a successful approach. One successful example is mentioned here. In 1975, Fu#66, an excellent mutant of wheat, was developed through γ ray (250 Gy) irradiation of hybrid seeds of Youbo/oro. Several fine cultivars relating directly or indirectly to Fu#66, Lumai#5 (1984), Lumai#8 (1985), Lumai#11 (1988) and more than 10 strains derived from Aimengniu/Fu#66. The cultivated area has covered more than one million hectares during the past five years.

As mentioned above, mutation breeding has been made remarkable progress in China. To improve mutation frequency and control the mutation direction are still the crucial issue on which the success or failure of mutation breeding depends. On the basis of experience of wheat mutation breeding of China, synthetic methods and techniques (Wang 1992), including the choice of initial stocks, use of various mutagens and new approaches for enhancing genetic diversity etc., have been studied and used in mutation breeding of wheat. Some successful examples of methods for developing mutant varieties and strains are listed in Table 10.6.

Table 10.6: Successful Examples of Strains or Varieties Obtained or Developed by Various Induced Methods

Names of Strains and Varieties Developed	*Different Methods for Developing Mutant Varieties or Strains*
Emai#6	γ Irradiation, pure variety [nanda2419]
Yuandong#3	γ Irradiation, hybrid seeds [12040/Auvror] F3
Hezu#8	γ Irradiation, [Zhe908 immature embryos]
Xifu#7	γ Chronic irradiation [Auror/Fan#7]
Chuanfu#1	B ray irradiation [Xifu#4]
Zhemai#3	CO_2 LASER 50J [E70]
Fushaibo#1	Fast neutron $6x10^{11}$ n/cm^2 [abo]
Aiyuansan	γ200Gy+ 0.3%EMS [Yuandong#3]
Wanyuan#28-37	γ350Gy+0.37%DES [st2422/464/Neixiang#5] F1
Wanmai#32	N^+ ion 30 Kev
85EA	Electron beam 370Gy[Yuandong2110]
Longfumai#4	γ110 Gy wheat/rye translocation strain6BS/6RL
Xiaoyan#6	Laser [st2422/Xiaoyue96]
89AR	Xinjiang xiaomai/g 20Gy [L. *raoo*]

Varietal Radio Sensitivity of Wheat (Feng and Wang, 1987)

The radio sensitivity of the initial material for wheat improvement plays an important role in determining mutation frequency. Radio sensitivity of wheat varied with species, varieties, tissues, cells, and different varieties based on the damage of morphological, cytological, and physiological and biochemical traits induced in M_1 generation. The varietal radio sensitivity of common wheat (*T. aestivum*) manifested a continuous variation, which accords approximately with the normal distribution, from the sensitive to the resistant to γ rays (Fig. 10.2).

By means of fuzzy clustering analysis which is based upon the Fuzzy Set Theory, wheat varieties were classified into five groups with different radio sensitivity, *i.e.* higher sensitive, sensitive, intermediate response, resistant and higher resistant. The experimental result has been found that the mutant varieties derived from radiation breeding were more resistant to γ rays than the local varieties, which are more resistant than the recombination varieties. In other words, cross–recombinant varieties > local varieties>mutant varieties in radio sensitivity. Spring wheats were sensitive than winter wheats. Radio sensitivity of wheat seeds was more resistant than tissue, cells and culture things etc. at various different developmental stages.

The radio sensitivity parameters of five different groups of varieties were showed in Table 10.7, and used in mutation breeding.

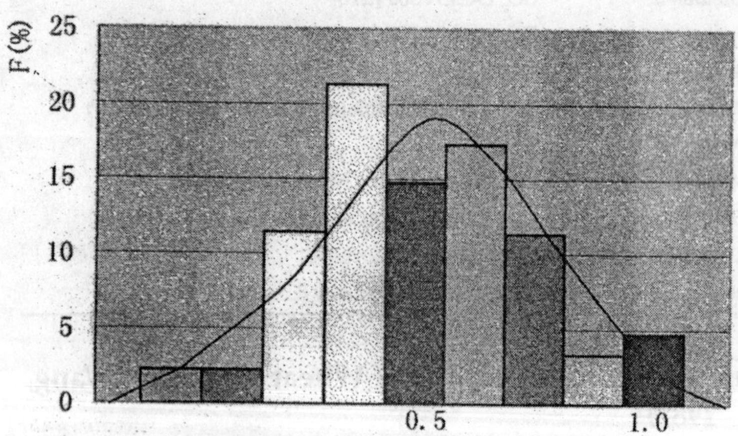

Fig. 10.2: The Distribution of the Varietal Radiosensitivity of Wheat (□ and F1 stand for □–level and the distribution frequency of wheat varieties, respectively)

Table 10.7: The Radio Sensitivity Parameters of Five Groups (Feng, Z.J. *et. al.*, 1987)

Groups	*λ-Level*	*HD_{50} (Gy)*	*LD_{50} (Gy)*	*YD_{50} (Gy)*	*RS (%)*	*$Vx10^{-2}$*
Higher sensitive	$0.10 < \lambda \le 0.28$	≤ 315	≤ 270	≤ 250	76-95	5.516-10
Sensitive	$0.28 < \lambda \le 0.46$	316-345	271-290	251-270	56-75	4.51-5.50
Intermediate	$0.46 < \lambda \le 0.64$	346-375	291-310	271-290	36-55	4.01-4.50
Resistant	$0.64 < \lambda \le 0.82$	376-405	311-330	291-310	16-35	3.51-4.00
Higher Resistant	$0.82 < \lambda \le 1.0$	> 405	> 330	> 311	< 15	2.90-3.50

Notes: HD_{50},LD_{50},YD_{50} stands for the doses that reduce seedling height, survival rate and seed yield by 50 per cent compare with the controls; RS stands for relative sensitivity and V for the coefficient of radio sensitivity

The disparity in radiation inhibition of DNA synthesis among varieties with different radio sensitivity had been studied, and it suggests that the reparability of DNA may be closely related with the varietal radio sensitivity. It is inferred that DNA reparability may be the main factor that governs the wheat varietal radio sensitivity.

Mutagenic effects of these varieties with different radio sensitivity were compared in M_2 generation. The results showed that there exists corresponding relationship between radio sensitivity and mutation frequency. Mutation frequency, mutation spectra of qualitative traits and variability of quantitative traits varied with varieties. Higher mutation frequency, wider mutation spectra and spectra variability have been observed in the extreme sensitive and sensitive groups than resistant ones (Fig. 10.3, Table 10.8).

The above results suggested that there is greater potential to select mutants in M_2 generation of radiosensitive varieties. Thus better mutagenic effect could be achieved if the radiosensitive varieties are chosen as the irradiated material in mutation breeding program.

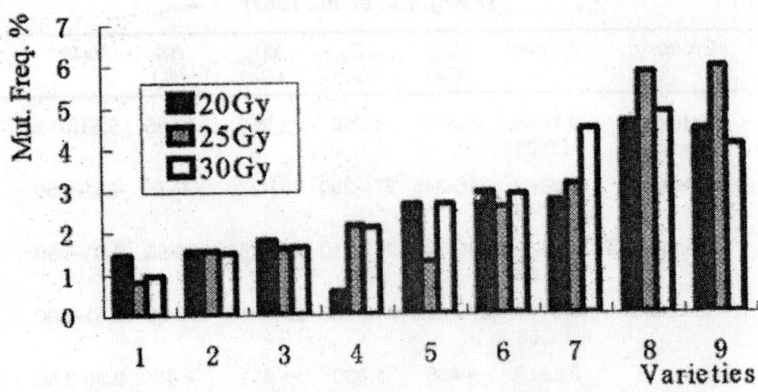

Fig. 10.3: The Relationship Between Radiosensitivity and Mutation Frequency.

1: Nongda311; 2: 1885; 3: March yellow; 4: Zhusanbo; 5: Xiaohongma; 6: Beijing15; 7: Beiling11; 8: Hongliang4; 9: 93Hong

Table 10.8: The Mutation Frequency of 4 Wheat Varieties with Different Radio Sensitivity in M_2 Generation (Feng, Z.J. *et al.*, 1987)

Radio Sensitivity Groups	*Varieties*	*γ-Doses (Gy)*	*Mutation Spectra and Frequency in M_2 Generations (%)*					*Total*	
			Early	*Dwarf*	*Spike*	*Awn*	*Others*	*Σ*	*%*
Sensitive	Hongliang	0 CK	0	0	0	0	0	0	
		200-300	0.45	1.68	1.97	0.64	0.16	4.9	48.76
Intermediate	Xiaohongma	0 CK	0	0	0	0	0	0	
		200-300	0.68	0.42	0.72	0.23	0.59	2.64	26.26
Resistant	Santuehuang	0 CK	0	0	0	0	0	0	
		200-300	0	0.36	1.12	0	0	1.48	14.73
Highly resistant	1885	0 CK	0	0	0	0	0	0	
		200-300	0	0.31	0.72	0	0	1.03	10.25
Total		0 CK	0	0	0	0	0	0	
		200-300	1013	1077	4.53	0.87	0.75	10.05	

Choice of Appropriate Initial Materials

The genotype and its radio sensitivity of the initial material play an important role in determining mutation frequency and mutation breeding. The achievements made in mutation breeding were closely related to the genetic background. The material can be seeds, embryoids, pollen grains, zygotes, and living plants at various stages and in vitro culture things, etc, depending on the specific needs. It is worth mentioning that mutation breeding can be used not only to induce mutation, but also to promote genetic recombination and to increase mutation frequency, expand mutation spectrum. Hence, pure lines or varieties as well as hybrid seeds, etc, can be used as initial stocks.

Irradiation of Hybrid Seeds (Xue *et al.*, 1989)

Irradiation of hybrid seeds of wheat increased the mutation frequency and fostered genetic recombination and widened the mutation spectrum; hence more mutants were provided for selection in breeding program. Of 91 mutant varieties of wheat, which were developed during 1966-1998, 42 (accounting for 46.2 per cent) were developed by direct use of mutants derived from irradiated hybrid seeds. Hybrid seeds used as initial material have been accepted in China as an efficient approach of mutation breeding in wheat, as well as in many other self–pollinated crops.

By studying the mutation frequency, including agronomic characters variation, chromosome aberrations, and isoenzyme variation and spectrum etc. in hybrid irradiated (F_1M_1) by g-rays and compared with parents (p) and their irradiated generation (PM_1) and hybrid F_1 without irradiation. Some interesting results have been found as follows:

1. Mutation Frequency of Traits Increased in F_2M_2. All of the plants expressed radiation injury degree of F_1M_1 was higher than that of PM_1 and varied range was F_1M_1>M_1 (Table 10.9). In other words, the radio sensitivity of hybrid F_1 was higher than that of parent varieties. Hybrids irradiated were much seriously damaged. The mutation frequency of main characters and spectrum of F_2M_2 was significantly higher and wider than those of both hybrid F_2 and parent M_2 (Fig. 10.4 and Table 10.9).

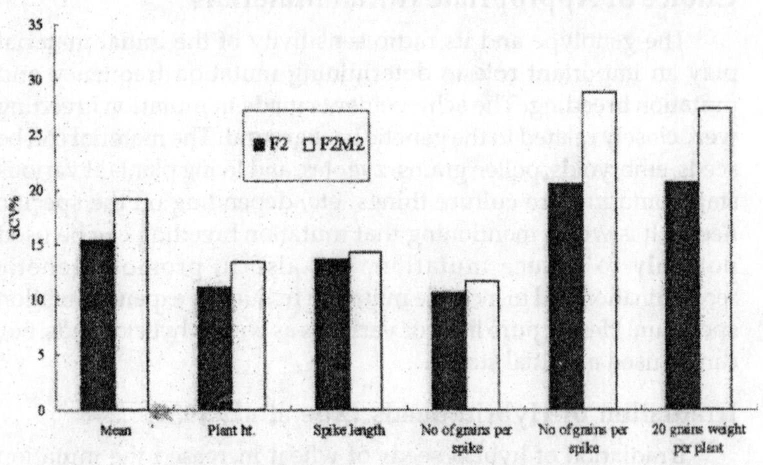

Fig. 10.4: Comparison of GVC (5 traits) Between F_2M_2 and F_2 (Wang, L.Q. *et al.*, 1985)

Table 10.9: Mutation Frequency (per cent) of Some Traits in F_2M_2 and F_2 (Wang, L. Q. *et al.*, 1989)

Generations	*Mutation Frequency of Traits*			
	Earliness	*Dwarfness*	*Spike Shape*	*Total*
F_2M_2	0.47	0.93	0.86	3.26
F_2	0.37	0.4	0.67	1.46
F_2M_2–F_2	0.08	0.53	1.19	1.8

2. The frequency of chromosome aberration at different division cell stages in mitosis and meiosis of F_2M_2 was not only higher than that of non-irradiated F_1 but also higher than that of parents irradiated (PM_1, Zhang 1995).

The frequency of dicentrics and chromosome fragments in metaphase of mitosis was higher than that of other types of aberration. The main chromosomal aberrations in metaphase of meiosis were fragments, monovalent and polyvalent chromosome bridges and laggings in anaphase of mitosis and meiosis. These

Table 10.10: The Chromosome Aberration of RTC and PM at Various Division Stages in Parents, Hybrids F_1 and Mutation Hybrid M_1F_1 (Xue, X.P. *et al.*, 1989)

Division Stages	*Treat-ments*	*RTC*							*PMC*						
		Parents			*Hybrids*				*Parents*			*Hybrids*			
		(1)	*(2)*	*(3)*	*(1)*	*(2)*	*(3)*	X^4	*(1)*	*(2)*	*(3)*	*(1)*	*(2)*	*(3)*	X^2
Inter-mitotic	CK	27112	10	0.05	25765	48	0.19	18.41							
	250Gy	20419	4059	19.87	14577	4169	28.80	359.6							
	X^2		4826.5**			8029									
Meta-phase	CK	1488	6	0.41	1861	13	0.70	1.24	401	1	0.25	419.29	0.92	26.1	
	250Gy	988	525	53.14	1094	830	75.89	117.99	896	401	44.75	730	573	78.49	442.2**
	X^2		957.3**			1908.4				256.9**			546.5**		
Anaphase	CK	1933	3	0.16	1213	25	2.06	30.3**	650	3	0.46	692	20	2.89	11.8**
	250Gy	1454	919	63.20	1141	927	81.24	10.13**	779	262	33.63	1124	629	55..96	92**
	X^2		1661.9**			1530.5**			258.1**				525**		

Note: (1) Total cells observed (2) No. Of cells with chromosome aberration (3) per cent of cells with chromosome aberration
**: Significant at 1 per cent level

chromosomal aberrations gave rise to chromosome translocation and exchange and the frequency of gene recombination was raised.

Table 10.11: Forms of Chromosome Aberration of RTC at Metaphase in P_1M_1, F_1, F_1M_1

Forms of Chromosome Aberration	*Parents*			*Hybrids*		X_2
	P_1	M_1	X_2	F_1	F_1M_1	
Observed cells No./%	1488	988		1861	1094	
Fragments No./%	5/3.5	234/23.68		6/0.32	376/34.37	
Micro body No./%	0	45/4.55		0	60/5.48	
Centric ring No./%	0	12/1.21		1/0.05	35/3.20	
Acentric ring No./%	0	20/2.02		1/0.05	21/1.92	
Dicentric body No./%	0	116/11.74		2/0.11	353/32.27	
Slimy lump No./%	1/0.07	53/5.36		2/0.11	56/5.12	
Total chromosome aberration No./%	6/0.41	480/48.58	580.9	12/0.65	901/82.36	2150.5
X^2				0.44	263.5	$X^2_{0.02}$=6.63

The frequency of micronuclei of intermeiosis cells, the chromosome aberration in metaphase and anaphase of mitosis and meiosis of the F_1M_1 were higher than the sum of the F_1 and P_1M_1.

From the results of these research studies, it could be concluded that irradiating hybrid seeds was an efficient way for wheat improvement.

Irradiation of Unicellular Cells-Zygotes and Gametes (Shi *et al.*, 1987, Zheng, *et al.*, 1985)

Dry seeds were subjected to irradiation. However, mutant cells are often inhibited by competition with normal cells in the polycellular formed seeds. Radios sensitivity of unicellular cells, such as zygotes, pollen grains etc., was greater than that of seeds and other cells. To minimize the diploid selection and the formation of chimerical structures, irradiation of unicellular zygotes is considered as one of the choices. Wheat breeders have used a more effective method–treating zygotes and gametes directly with mutagens. Some experimental results showed that after irradiation of gametes and zygotes of wheat by γ ray with appropriate dose, a high mutation frequency, wide spectrum and uniform mutants could be obtained. Judging from the mutation rate and spectrum, this treatment was more effective than seed treatment.

The mutation rate and spectrum of zygotes irradiated varied with the developmental stages of zygotes. The cell cycle of zygote stage of wheat was determined by means of microautoradiography with ^{3}H-TdR as a tracer. According to the observation, analysis of micronuclei rates and chromosomal aberration rates, a peak of radio sensitivity was detected at the DNA synthesis phase or early G_2 phase. The period of DNA synthesis range from 8 to 10 hours after pollination. The highest mutation frequency was obtained at this stage (Fig. 10.5, Table 10.12).

Table 10.12: The Fertilization and Duration of Zygote of Wheat (Beijing)

Material		*Stage of Fertilization After Pollination (h)*		*Zygote Period [After Pollination (h)'*	*First Mitosis of zygote (After Pollination)*
		Pollination to Enter embryo	*Nuclear Fusion to Zygote Stage*		
Wheats	Winter	0.5	7-Apr	13-Jul	14-17
	Semi winter	0.5	6-Jan	11-Jun	15-Dec
	Spring	0.5	7-Jan	23-Jul	24

In general, a damage effect appeared and wide variation produced which could be transferred in M_1 generation, the macro-mutation did not produce in M_2 generation only negligible appeared in a few characters. More than 60 per cent mutant line with stable traits could be obtained in M_3 generation, thus simplified the breeding processes.

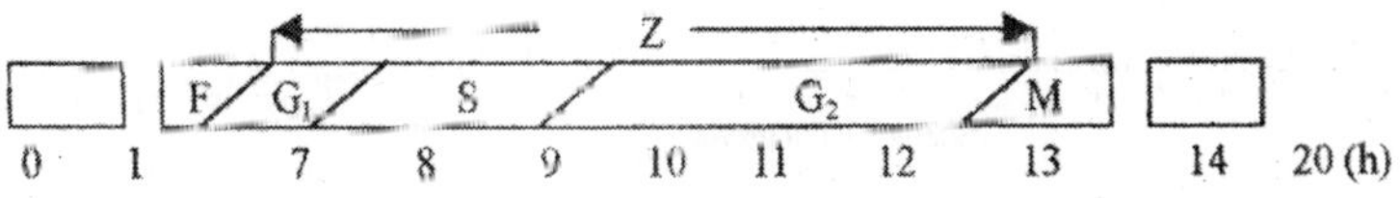

Fig. 10.5: Stages of Pollination Fertilization and Zygote of Wheat

Table 10.13: Chromosomal Aberration % and Micronuclei % After Zygote Irradiated of Wheat (Zhang B.L. *et. al.*, 1989).

Irradiating Time (After Pollinating)	*Cycle Phase*	*Chromosomal Aberration (%)*		*Micro-nuclei (%)*	
		RTC (1) (M_1) (%)	*PMC (2) (M_1) (%)*	*RTC (3) (M_1) (%)*	*PMC (4) (M_1) (%)*
7	G_1	11.2	7.0	2.9	4.0
9	S	33.0	20.6	4.0	8.3
11	Early G_2	25.0	17.0	14.0	17.0
13	AfterG_2	12.0	11.3	6.0	14.0
CK		4.8	2.8	0.3	1.0

Note: (1) No of RTC observed, 206-241. (2): No. of PMC observed 1002-1259. (3): No of RTC observed 1003-1066. (4): No of PMC observed 1047-1147.

The zygotes of wheat were very susceptible to irradiation, which by inhibiting the development of embryo and endosperm caused germination failure of seed and a relatively low survival rate of M_1 seedlings. These factors undoubtedly affect in M_2 population.

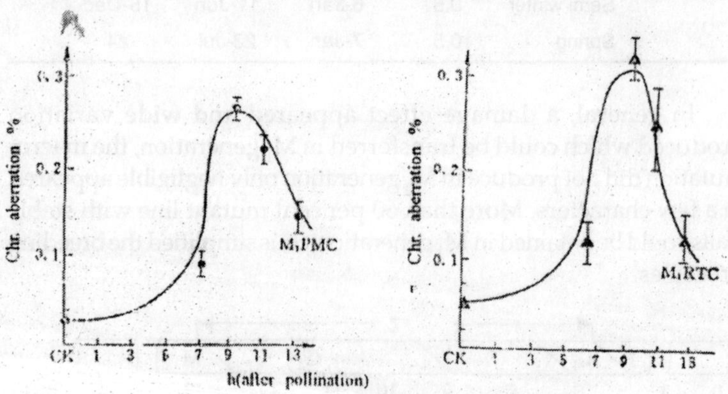

Fig. 10.6: Chromosomal Aberration Frequency (per cent) Observed in M Plants of Wheat Left: PMC, Right: RTC.

However, zygotes were irradiated with appropriate doses at the proper time, and by supplementing tissue culture technique. It was advantage to obtain more mutant material. Some favorable mutants, such as disease resistance, Nacl tolerance and high protein content mutants etc., have been obtained and used in breeding program. Some of them have been developed new varieties and cultivated for production.

Based on some experimental results, male and female gametes treatment resulted in a lower chromosome aberration rate in M_1 generation and a higher mutation frequency and wider spectrum in the M_2 generation. The favorable mutation frequency was higher than that of seed treatment. This method was easier to apply than that of the zygote treatment. It is considered to be a method of mutagenic treatment of great practical value and some desirable mutant strains have been developed and used in breeding program and cultivation.

Irradiation of Explants, Cultured Tissue and Cells (Wang 1998)

Using explants, such as young spikes, embryos, and leaves etc., as well as the anthers and haploid cells in culture irradiated with appropriate doses and in vitro culture could be increased the mutation frequency and expanded the mutation spectrum, without forming chimeras, minimizing diploid selection and the mutant traits (or characters) at the cellular level could be identified (or detected) easily. Induced mutation combined with bio-technique has been used widely in wheat breeding program. Anthers were irradiated with g-ray before inoculation. Phenotypic variation in the plants derived from irradiated anthers was found to be higher than that of non–irradiated anthers. Most variants found in γH_1 were bred true. Some were valuable mutants such as, Nacl-tolerant mutant. A number of Nacl-tolerant cell line were selected from a great quantitative of callous cultures. The Nacl-tolerant genetic mutants of 5 lines were stable and very strong after subculture of three times and one cultivar $H_{92\text{-}112}$ have been developed and released to farmers. Another mutant cultivar, Hezu#8, with early maturity, main disease resistance, wet tolerance, good quality and high yield derived from Zhefu 908 immature embryo irradiated by g-ray. The experimental results showed that 60Co g-ray was used to induce mutation and 10 Gy was found to be effective.

Use of Various Mutagens (Liang and Wang 1987)

Various physical mutagens such as γ ray, β ray, neutrons, including fast neutrons, thermal neutrons, lasers, electron beam, ion beam and artificial satellite etc., and some chemical mutagens, such as ethylmethanesulphonate (EMS), diethyl sulphate (dES), Ethylene amine (EI), N-Nitroso-N-Ethylurea (NEU) and Sodium azide (NaN_3) etc., have been used in mutation breeding either alone or in combination with each other. The number of mutant varieties developed by the direct use of mutants was showed in Table 10.14.

Table 10.14: Number of Released Mutant Varieties Derived from Different Mutagens of Wheat in China (Wang 1998)

Years	*Mutagens*						*Total*	
	Gamma Ray	*Beta Ray*	*Neu-trons*	*Lasers*	*Ion*	*Mutagens combined*	*No*	*%*
1966-1980	38	—	1	—	—	1	41	45.0
1981-1990	18	1	3	1	—	2	30	33.0
1991-1998	15	1	1	6	2	1	20	22.0
Total No	71	2	5	7	2	4	91	100.0
%	78.0	2.2	5.5	7.7	2.2	4.4	—	—

γ Rays were the most widely used mutagens, and 71 (accounting for 78.0 per cent) of the mutant varieties were developed after γ ray irradiation. 20 out of 91 mutant varieties were developed by other mutagens. Two mutant varieties, Taifu # 15 and Zhengliufu, developed through replicated irradiation by γ ray, Xifu # 7 from γ ray chronic irradiation in the γ-field located in Sichuan province, China.

Neutrons, fast neutrons and thermal neutrons, produce higher ionization density and strongly ionizing radiation. The inducing effect of neutron was higher than that of X-ray and g-ray. It was also less affected by environmental factors. Some desirable mutants and five mutant varieties, such as, Longfumai#1, Fushiabo#1, and Xifu#7 which was 10-15 days earlier, stem height 10-15 cm dwarf and better rusts resistance than the initial material, were developed by tast neutron irradiation. However, this method has not yet been used widely in crops as well as in wheat, since it was expensive and the absorption dose was not easy to measure.

Along with the developments of acceleration techniques, space techniques, and some other new techniques, some new mutagens, including Beta-rays, lasers (a Co_2 laser, a He-Ne laser, N_2 laser and Nd glass laser etc)., ion beam (H^+, $N^+$and Ar^+ etc). implanation, electron beam, proton etc., have been exploited, studied for their mutagenicity as well as biological effect and used for improving genetic diversity and raising the level of mutation breeding of wheat. Some primary experimental results and useful mutants used in breeding program have been obtained and 11 mutant varieties (accounting for 12.1 per cent) have been developed and released to farmers for production. For successful examples: Xiaoyuan#6 with rusts and scab resistance, good quality, wide adaptation and high yield was derived from (st2422/Xiaoyan96) treated with Nd glass laser, Zhemai# 4 and Lumai# 16 were developed by Co_2 laser and He-Ne laser treatment, respectively. Two varieties Chuanfu#1 and Longfumai#5, developed through Beta-ray irradiation, N^+ ion beam implantation from ion accelerator with various appropriate dose (about 10^{14}-10^{16} N^{+}/cm^2) as well as protons, electron beam etc., were able to generate variations on the chromosome structure. The rates of cells with chromosome aberration were increased with increase of ion beam energy and dose, but the dose effect of ion beam was higher than that of energy. As compared with gamma ray, various micro mutation and point mutation were higher gamma ray irradiating. Those mutagens could be effective new mutagens for wheat improvement. Some desirable mutants and good mutant varieties have been developed and used in breeding program or released for production. Cytoplasmic male sterile line, CMS 85 EA, was derived from electron beam (5 Mev linear accelerator) with 370Gy. Its fertility was easier to recover and possessed wider restoring resources than *T. timopheevi* CMS and have been used to wheat heterosis breeding program. New cultivars, Longfumai#8 and Wanmai#2 with good plant type and synthetic agronomic character were developed by N^+ion (30 Kev) implantation in 1996 and 1997, respectively.

About space breeding was practiced in China in recent years, using Near-ground space treatment (NGS) and artificial satellite, NGS possessed stimulating effect on the growth of seedling compared with 300 Gy of gamma ray irradiating, the damage effect of NGS on M_1 generation were reduced in some crops, as well as in wheat. But, some results showed that there were more chance to generate good variation in M_1 by NGS treatment and the variation rate were raised. Some interesting mutants, including CMS, restoring (Rf) line and large grains etc, have been obtained and methods for improving mutation frequency were being studied. However, many problems of using those new mutagen factors mentioned above have to study and discussed.

The combined use of physical and chemical mutagens has been an effective way for raising the mutation frequency and expanding the mutation spectrum. Some studies and results (LI *et al.*, 1989, Sun *et al.*, 1995) showed that additive effects as well as synergistic effects were observed in the M_1 generation treated with gamma ray + EMS, gamma ray+dES and gamma ray +NaN_3 etc. The biological injuries off M_1 generation increased with the dose of gamma ray and the concentration of chemical mutagens,. The mutation frequency and mutation spectrum of combined treatment of M_2 generation were higher than those of treatment alone were. The highest mutation frequency were observed in the combined treatment of c + 0.3 per cent EMS and 200 Gy g ray+ 2mM NaN_2 (Fig. 10.7). It could be recommended for mutation breeding of winter wheat. The mutation frequency of the M_2 generation was closely related to the seedling height, vitality index, root length and percentage of root tip cells with single micronuclei of M_1 in both combination treatment, especially the single micronuclei of RTC in 200 Gy g ray irradiation

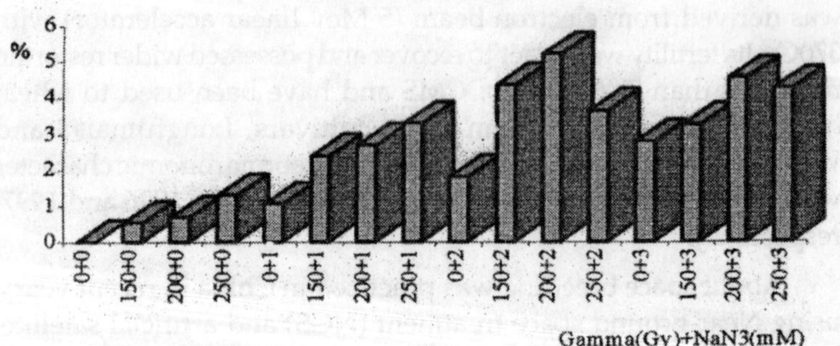

Fig. 10.7: Distribution of the Frequency of Favorable Mutations in M_2 Generation Induced by Gamma Ray Combined with NaN_3

and combined with 0.3 per cent EMS treated combination and others in g ray 200 Gy + 2mM NaN_2 combination treatment (Table 10.15)

Table 10.15: Biological Criteria of M_1 in the Best Rreatment (200 Gy g ray + 2mM NaN_3) for Predicting Mutagenic Effect in M_2 (Li *et al.*, 1989)

Varieties	*Treatment*	*Seedling Height*	*Root Length*	*Index of Vitality*	*Mono-micro-nuclei*
Yuandong#3	CK	17.4	13.9	326.27	
	200 Gy + 2mM NaN_3	9.3	7.1	71.1	6.4
	200 Gy + 2mM NaN_3/CK	0.53	0.51	0.22	
Nongda146	CK	13.5	12	197.53	
	200 Gy + 2mM NaN_3	8.1	5.6	52.08	6.5
	200 Gy + 2mM NaN_3/CK	0.6	0.47	0.26	

The results showed that one half of seedling height and root length, one fourth of the vitality index of control (CK) and 6.25 per cent cells with single micronuclei in M_1 could be used for predicting mutagenic effect of M_2 at an earlier stage under the appropriate combination treatment of physical and chemical mutagens. Some desirable mutants such as powdery mildew and rusts resistance, dwarfism, early maturity and higher yield etc., have been obtained through this method and used in cross breeding. One of the outstanding mutant cultivars, Aiyuanshan derived from gamma ray 200Gy+ 0.2 per cent EMS [Yuandong#3] in 1996, possess semi-dwarf stem, 15 cm shorter than the original one, rusts and powdery mildew resistance, drought and saline alkali tolerance, higher yield and good wider adaptation, have been released for commercial production.

Identification and Selection of Mutants

The methods of identification and selection of mutants also affected the efficiency of mutation breeding. The mutation frequency of various induced traits was very low, especially disease resistance and tolerance to the environment stresses. To choose appropriate initial material, a large population of mutation progenies and general method of individual selection were both important and necessary. Recently, newly developed methods of identification and selection, using in vitro culture and biotechnique such as, RFLP, RAPD etc., have been achieved in this field. Several disease resistant mutants and environmental stresses tolerant mutants have been obtained. For Example: the mutant strains of spring wheat with root rot (*B. sorokiniana*) resistance (Sun *et al.*, 1989) have been obtained through induction mutation and embryo culture in vitro which rough toxin of *B. sorokiniana* was added to the differentiation medium used as the screening factor. Nacl tolerant strains of wheat were selected and by using in vitro culture technique (Xiao *et al.*, 1989). Some mutants with good physiological traits and disease resistance were identified by using isoenzyme technique, such as the patterns of peroxidase and esterase isoenzyme etc., were used to identify powdery mildew resistance in wheat. The inheritable behavior of mutant traits has been studied using cytological and genetic methods (Sun *et al.*, 1995).

Application Developed of Induction Mutation

Chromosome Translocation for Creating New Gremplasms

Chromosome translocation induced by exploiting radiation has been studied and achievements made in wheat and its relative (Li *et al.*, 1985). By inducing chromosomal translocation alien genes or chromosomal fragments can be incorporated into receptor species. The procedure for inducing, identifying and screening chromosomal translocation strains has been established in wheat breeding. Some new translocation strains and varieties of wheat with rye characteristics have obtained and developed. The Longfumai#4, a new cultivar of spring wheat, derived from the 6BS/6RL translocation line, has been cultivated in Heilongjiang province. Northeastern China. Chromosomal translocation techniques are widely applied in wheat mutation breeding program.

Irradiation for Overcoming Incompatibility of Wide-Cross and Alien Gene Transfer

As is well known, a lot of desirable genes exist in wheat related species and genus. By means of wide cross, alien genes of widely related species could be introduced into the genotype of wheat varieties or strains. However, the problems of obtaining distant hybrid seeds were not easy to solve. Some experimental results

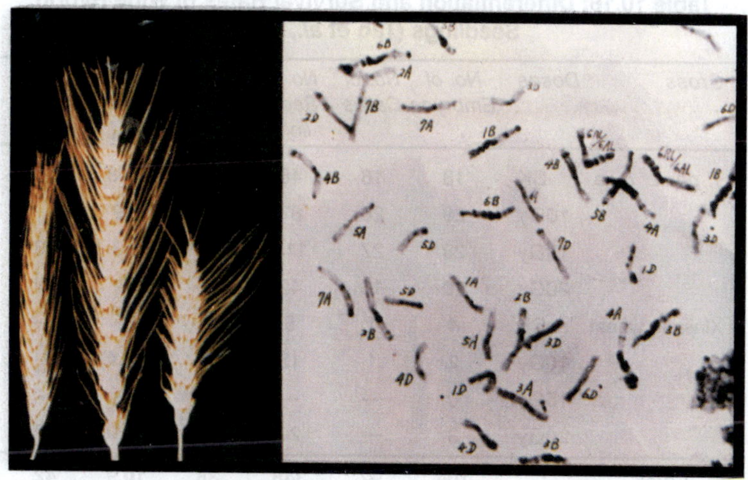

Fig. 10.8: Chromosomal Translocation Lines (6RL/6AL) Left: Rye, Med.: Triticale, Right: Wheat.

showed that it was possible to enhance wide cross by using an appropriate low dose or chronic radiation and hybrid embryo rescue in vitro culture. Some interesting results have been obtained over the past 10 or more years. For example:

1. The pollen grains of *Leymus racemosus* (2n=28) were irradiated by g ray with low doses (10Gy, 15Gy and 20Gy), and crossed with *T. aestivum*, Xingjiang wheat (2n=42), and *T. persicum* (2n=28, Tao *et al.*, 1995, Table 16). The immature embryos cultured in vitro were used to subculture and propagation. The percentage of seed set was increased, and acquired rate of progeny of wide cross was rising substantially. Some hybrid plants with both wheat and L. racemosus traits of F_1M_1, F_2M_2 and F_3M_3 were obtained and studied in morphological, cytological characters and patterns of peroxidase isoenzyme (Tables 10.16 and 10.17). One hybrid plant, male sterile CMS 89 AR was appeared, developed and used in heterosis of wheat.

Table 10.16: Differentiation and Survival Rates of Wide Hybrid Seedlings (Tao *et al.*, 1995)

Cross	*Doses*	*No. of Embryos*	*No. of Callus*	*No. of Seedlings*	*Survival Seedlings*	*Survival Seedlings (%)*	*No. of Hybrid Plants*
T. per./L. race.	CK	18	18	48	10	20.8	—
	10Gy	29	26	87	6	6.9	—
	15Gy	28	27	114	6	5.3	13
	20Gy	20	16	42	0	—	—
Xinjiang wheat	CK	4	1	5	0	—	—
	10Gy	2	1	13	2	15.4	—
	15Gy	0	—	—	—	—	—
	20Gy	3	—	29	14	35.9	29
Total		**104**	**92**	**348**	**38**	**10.9**	**42**

Table 10.17: Comparison of Characters Between Parents and Hybrid Plants

Characters	*Xinjiang Wheat*	*Hybrid Plants*	*L. racemosus*
Pt. Ht. (cm)	120	120-163	120-140
Stem diameter (cm)	2.2	3.5	6
Leaf length (cm)	20	30	75
Leaf color	Green	Dark green	Blue green
No. of spikes/pt.	Several	73	many
Spike type	Spindle	Sharp tower	tower
No of spike lets/spike	Awn less	top awned	awned
Isozyme peroxidase		one new spectrum	

2. Enhancement of intergeneric hybridization between common wheat (2n=42) and Leymus angustus by means of nuclear irradiation (Wang 1998). Choosing for suitable wheat genotype, J-11, determining of appropriate irradiation low dose "9Gy" of gamma ray, the pollens of L. Angustus irradiated with 9Gy of gamma ray and was crossed with J-11. The immature hybrid embryos were

rescued with "embryo-callus culture methods". The embryo acquired rate, callus induction rate, percentages of plantlets and survival of hybrid seedlings were raised (Table 10.18). New plants could be obtained.

Table 10.18: Enhancement of Intergeneric Hybridization Between Common Wheat and *L. ang.* by Means of Irradiation (Li *et al.*, 1999)

Combination	*Embryo Acquired (%)*	*Callus Induction (%)*	*Plants From Culture (%)*	*Survival Plant (After Transplant) (%)*
J-11/*L. ang.* (CK)	1.24	33.33	18.61	3.62
J-11/*L. ang.* (9Gy)	3.79	41.67	27.21	89.4
J-11/*L. ang.* (9Gy)-CK	2.55	8.34	9.60	85.74

This method mentioned above, irradiation with appropriate low dose of g ray, would be an effective method for overcoming incompatibility of wide cross and enhancing alien gene transfer. Some evidences were acquired on hybrid plants, obtained, chromosome behavior (Fig. 10.9) and genomic in situ hybridization analysis and etc.

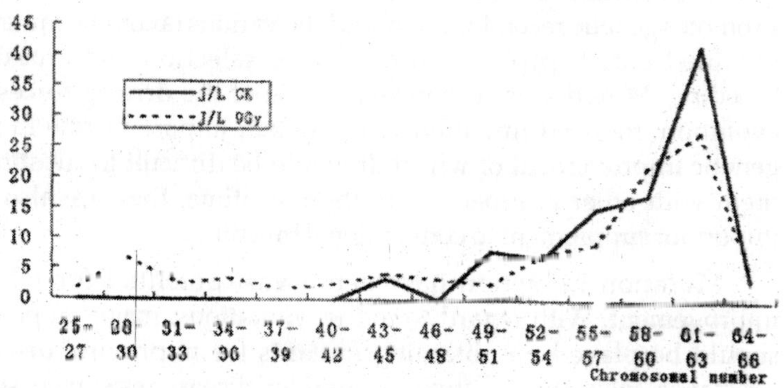

The methodology, mentioned above, would improve the efficiency and level of crop mutation breeding.

3. γ Ray chronic irradiation for enhancing wide cross (Chen *et al.*, 1995)

The effects of ^{60}Co γ ray chronic irradiation on intergeneric cross between wheat and barley conducted in recent years. The male and female parents were planted in the γ-field and irradiated with appropriate low dose of γ ray at differentiation at differentiation stage of sex organs of wheat and barley. After the hybridization has been done and hybrid embryos cultured in vitro. The results showed that the hybridization rate of crossing between wheat and barley was increased. The hybrid plants F_1 [2n=28], [7 from barley (2n=14) and 21 from Chinese Spring (2n=42)], were obtained from hybrid embryos. The progenies of hybrids (F_1, F_2, BC_1, BC_2) could be set seeds by self-crossing. The variation and segregation in seed setting percentage, plant height, spike length, number of spike lets and floscular, spike form. awedness, grain shape, grain color, grain weight, and plumpness etc., were very rich and obviously surpassed the of varietal crossing from non-irradiation. Some desirable variation plants have been obtained in intergeneric hybrid progenies.

Outlook

Use of induced mutation generates greater genetic variability, promotes genetic recombination, and the various favorable mutants obtained enrich plant gremplasms for selection and crossing breeding. Mutation is recognized as one of the driving forces of evolution. Induced mutation can play ban important role in the genetic improvement of wheat. It would be difficult to substitute them with other approaches. In the meantime, they are also an important supplement to conventional breeding.

Mutation induction has been a vast potential for wheat improvement. With regard to induce mutations, major emphasis should be played on obtaining mutants for improving disease resistance, including multiple or complex disease resistance, such as rusts, powdery mildew, scub, root-rot and dwarf virus etc., improved tolerance to environmental stresses, especially drought and saline-alkali in North China and other places. Some mutants with special traits, such as male sterility and restorability and etc., for other special needs in breeding program and production.

Application of the mutagenic technique to in vitro culture biotechnique and molecular technique has attracted increasing interesting over the recent past years. Those techniques have been used to raise the efficiency of the mutation induction and selection of mutants. Combined use of induction mutation, biotechnique and molecular techniques could enhance wide cross and transfer the desirable alien genes from relative species or genus to wheat and create new valuable gremplasms for wheat improvement.

To improve the level of mutation breeding and promote its development, attention should be paid to basic research and a combination of mutation breeding and other approaches of wheat improvement.

REFERENCES

An,D.C., Wang,L.Q. 1985. Mutagenic effects of electron beam, neutron and (ray in common wheat, Appl. At. Energy Agric. Suppl.issue, p104-109.

Chen, Y.L. *et al.*, 1995. Effect on chronic irradiation on intergeneric cross between barley and wheat, In: Induced mutation for crop breeding. At, Energy press. p436-446.

FengZ.J., Wang, L.Q. 1987. Varietal radio sensitivity and mutagenic effects in Triticum aestivum L. Report on Nuclear Sciences and techniques in China. CNIC-00066,1987, p1-16.

Jin, S.B. 1996. Wheat of China. China Agric. Press, P4, P257.

LI, S.R. *et al.*, 1989. Mutagenic effects of 60Co (ray combined with sodium azide in winter wheat, Acta Agric. Nucl. Sinica,. Suppl. Issue. p155-164.

Li,J.L. *et al.*, 1985. A study on induced chromosome translocation by exploiting ionizing radiation in Triticale, Appl. At. Energy Agric. Suppl, issue. p64-74.

Liang, Q., Wang, L.Q. 1987. Mutagenic effects of 60Co (ray combined with EMS in wheat. Report on Nuclear Sciences and Techniques in China. p1-14.

Micke,A., *et al.*, 1987. Induced mutation for crop improvement–a review, Trop Agric. (Trinidad), Vol, 4,Oct, 1987, p259-278.

Savascia-Mungnozza, G.T.et al. 1991. Mutation breeding programme for durum wheat improvement in Italy, In: Plant mutation breeding for crop improvement. Vol 1, IAEA. p95-109.

Shen, S.J. 1995. Progress in studies of mutagenic agents for mutation breeding, In: Induced mutation for crop breedin. At, Energy press. p187-191.

Shi,J.G. *et al.*, 1987. Inducing mutation by treating male gametes and zygotes of winter wheat with 60Co (ray and EMS, Appl.At.Energy in Agric. 2: 1-6.

Sun, G.J. *et al.*, 1995. Studies on location gene of resistance to yellow rust and dwarf in wheat mutant Yuandong 96, In: Induced mutation for crop breeding. At, Energy press. p345-351.

Sun, G.Z. *et al*, 1989. Screening for mutants of resistance to B. sorokiniana by irradiation and in vitro technique. Acta Agric.Nucl. Suppl,issue. p177-182.

Tao, S. H. *et al.*, 1995. Studies on overcoming incompatibility of wide cross by irradiation and enhancing alien character transfer. In: Induced mutation for crop breeding. At, Energy press. p447-456.

Wang, L.Q. 1991. Induced mutation for crop mutation breeding in China, In: Plant mutation breeding for crop improvement. Vlo.1, IAEA, p1-32.

Wang, L.Q. 1992. Synthetical techniques for crop improvement through radiating breeding, Sciences & Techniques. 11: 24-26, 12.

Wang, L.Q. 1995. Advances and achievements of crop mutation breeding in China. Induced mutation for crop improvement. At. Energy Press,,p1-11.

Wang, L.Q. 1995. In: Induced mutation for crop breeding. At, Energy press. p1-456.

Wang, L.Q. 1996. Induced mutation breeding of wheat. In: Wheat of China. p381-401.

Wang, L.Q. 1998. Methodology of enhancement genetic diversity for crop improvement by induced mutation in China, In: Methodology for plant mutation breeding. Malayasia. p1-8.

Wang, L.Q., *et al.*, 1996. Study on the Cytoplasmic male sterility (CMS) lines of 85EA and 89AR in common wheat. In: Recent advances in crop male sterility and heterosis. Vol., I, China Agric. Press, p136-144.

Wang, L.Q.,et al. 1985. Studies on improving the efficiency for inducing mutation of wheat hybrid by (irradiation, Appl.at. Energy in Agric. Suppl.issue, p39-44.

Xiao, H.L. *et al.*, 1989. The selection of Nacl-tolerance mutants by tissue culture in wheat, Acta Agric. Nucl. Sinica, 3(2): 85-90.

Xu, G.R., Wang, L.Q. *et al.*, 1996. In: Induced mutation breeding of plants. China Agric press. p1-492.

Xue, X.P. *et al.*, 1989. Studies on improving the efficiency for inducing mutation of wheat hybrid by (irradiation, Acta Agric. Nucl. Sinica. Suppl, issue, p135-147.

Zhang, B.L. 1995. Zygotes phases of crop with induced mutations, In: Induced mutation for crop breeding. At, Energy press. p133-138.

Zheng, Q.C. 1995. Advances and application of induced mutation breeding technique in vitro, In: Induced mutation for crop breeding. At, Energy press. p385-392.

Zheng, Q.C. *et al.*, 1985. Induced mutation in anther culture of wheat, Appl. At. Energy Agric. Suppl. Issue, p170-184.

Chapter 11

Selection Studies on Mutant Barley Populations

M.B. Yildrim, N. Budak, Z. Yildrim & T. Kusaksiz

Ege University, Faculty of Agriculture, Department of Field Crops, Bornova-Izmir, Turkey

ABSTRACT

Six mutant hybrid barley populations constructed by crossing Quantum and Kaya mutants and 4 control varieties were grown at two locations (at Bornova and Menemen) in the 1992-1993 and 1993-1994 growing season.

A total of 400 single plants were selected at each mutant and control populations in 1992-1993 growing season. Ten percent selection pressure was applied and 40 mutant lines from each population was selected. The progeny rows of the selected mutant plants were grown at two location in 1993-1994 growing season. A second stage selection was applied in each progeny populations and the selected mutants were advanced to the micro yield testing. Doing each stage of selection genetic gains for single plant yield and expected progeny means for plot yield were estimated.

Mutant lines with high single plant yield were found in certain mutant populations. The progeny means of the selected mutant populations were higher in mutant populations. And the expected genetic gains were also found to be high for the mutant selections.

INTRODUCTION

Barley is consumed as food, and malt. Several by-products of malt are also used in human consumption. The feed value of barley is in good standing among cereals. In Turkey, barley is grown at about 3.5 million hectars area with a production of 7.5 million tons. Developments in livestock raising and in malting industry make barley more important. Turkey has a production capacity of 150 thousand tons for malt. Malting industry demands about 200 thousand tons of raw material. On the other hand, barley suitable to malting is not grown in Turkey. Therefore barley cultivars with accepted malting characteristics such as Tokak and Kaya are grown. Recently some new varieties such as Efes types were also released by a malting company. Tokak, two rowed barley, has been widely grown for several years. In the Aegean Region Quantum was released by us and Kaya was released by the Agricultural Research Station Institute in 1976. Combination breeding has been used in barley breeding but now varieties are mainly the results of selection applied in the landraces (Tokak) and in certain foreign introductions (Quantum and Kaya).

Combination breeding creates new variation to be used in further selection but it requires suitable parents, several crossings and large amounts of the F2 materials beforc applying selection. Therefore plant breeders have used induced mutation in order to obtain large genetical variations for desirable traits. Beside, mutation breeding could be conducted in a more simple and suitable way in breeding new barley lines. Therefore, combination breeding and mutation breeding could be utilised separately or even jointly by barley breeders (Gaul 1963, Akbay 1977, Yýldýrým and Tugay 1977).

We have started mutation breeding at the Aegean University with financial supports from the Turkish Atomic Energy Commission in 1973. At this time we had onc malting barley variety selected and proposed to released by us (Quantum) and one feed barley (Kaya) released by the Regional Research Institute. Thercfore we have selected these varieties as starting material.

Yýldýrým *et al.* (1987) reported the results of the single plant selection applied in the M2 generation of irradiated Quantum and Kaya populations. Çaðýrgan and Yýldýrým (1988) reported barley mutants selected in Kaya and Quantum barley populations irradiated by gamma rays at 150 Gy such as Anthocynadin-free, erectoid, early heading and male sterile and discussed their possible utilization in barley breeding.

Çaðýrgan (1989) has studied the M_2 and M_3 generations of mutant populations in terms of quantitative trends and compared single plant yield with plant yield basis selections.

Çaðýrgan and Yýldýrým (1990) selected single mutant plants in a Kaya populations irradiated by gamma rays (150 Gy). They found that there were large variations for several traits as compared to the control populations. Although mutants such as early, short and with high thousand-kernel weight were found, there was not any high yielding than the control among the selected mutants.

Yýldýrým and Budak (1990) studied the progenies of single mutant selections. They found significant differences among mutant progeny rows in terms of plant height, earliness, tiller number, thousand kernel weight and spike length. They selected high yielding 19 mutant lines as compared with the control. The heritability of yield was 0.69.

Yýldýrým and Çaðýrgan (1991) selected randomly individual plants in M_2 population of Kaya variety irradiated with gamma ray. They selected 25 per cent of the upper yielding plants for progeny testing. They applied second selection pressure in the progeny testing stage or yield. They reported that higher yielding mutant plants and progeny rows originated from the base population.

Molina-Cano (1982) selected individuals plants in the M_2 generation of Beka variety and tested them as progeny rows. At the end, they found high yielding mutant lines. Garcia del Moral *et al.* (1991) crossed mutant lines obtained in Beka barley and compared them with Beka and Zaida varieties. They reported that the crosses between mutants were higher than their parents and the control in terms of spike number, grain number per spike and grain yield.

Efe and Yýldýrým (1992) studied 17 mutant lines of Kaya variety. They reported that mutant lines exceeded Kaya for thousand-kernel weight, spike length, flag leaf length and grain number per spike. They concluded that there was a possibility increasing via yield components in mutant lines.

The purpose of this study was to investigate the effects of a second stage selection on the certain barley populations constructed by crossing mutant lines in terms of yield and estimate the genetic gains

Materials and Methods

Six mutant hybrid barley populations and 4 controls used in the study are given in Table 11.1.

Table 11.1: Barley Mutant Hybrid Populations and Controls Used in the Study

Population No.	*Specific Characteristics*
POP 1	Base population established by crossing mutant lines of Kaya and Quantum.
AP 1	Constructed by selecting the genotypes with high spike yield in the base population.
AP 2	Constructed by selecting the genotypes with high kernel number in spike in the base population.
AP 3	Constructed by selecting the genotypes with high thousand-kernel weight in the base population.
POP Q	Male sterile population of Quantum. Constructed by mixing the male sterile genotypes selected in the irradiated Quantum.
POP K	It is the male sterile population of Kaya mutant lines.
QUANTUM	Control variety; two rowed high yielding, spring type and good beer quality.
KAYA (I-9)	Control variety with good malting quality, spring type, two rowed also suitable to animal feeding.
TOKAK	Control variety, spring type, two rowed, resistant to winter hardiness and good for beer and animal feeding.
EFES	Control variety, spring type, two rowed, resistant to winter hardiness.

Six mutant cross populations and 4 control populations were grown at Bornova and Menemen locations in the 1992-1993 and the 1993-1994 growing season in order to estimate the heritability and conduct selection work. The design of the experiment was a Randomized Complete Blocks with 4 replications. One plot consisted of 6 rows with 5 m in length and 20 cm between rows. Planting were done in December in both seasons with machinery. Regular agronomic applications were applied and 6 kg P205 and 8 kg N were given at each location. In the 1992-1993 growing season, 400 individual plants were selected randomly from each population at each location in order to construct the base populations. Yield and other agronomic traits were measured on each single plant. The 10 per cent high yielding individual plants were selected from each base population. The same procedure applied to the control populations. Forty individual plants from each population were grown as progeny rows in 1-m single row plots with 30 cm spacings. Planting was done on December 18 at Bornova and December 23 at Menemen in 1993. Harvesting was completed in first week of July at two locations in 1994.

Agronomical traits such as plant height, spike length, number of grains per spike, heading time, maturity time, harvest index, plot yield and thousand kernel weight were recorded at each location. The protein contents of progeny were determined by using the "Near Infrared Reflectance Spectroscopy (NIR-S)" technique (Thachuk 1981).

Statistical Analysis

The data obtained from two years growing of original populations at Bornova and Menemen were analysed and combined over years and locations in order to get dependable heritability estimate by applying variance components method (Allard 1960).

Single plant yields of the selections done in the original populations were analysed statistically assuming normal distribution and mean standard deviation and the coefficient of variation (C.V). values were calculated (Steel and Torrie, 1981). The genetic gains were estimated by the following formula described by Allard (1960).

Genetic Gain=H. Sf. I

Where; H= heritability in broad sense

Sf= Phenotypic standard deviation

i= selection intensity (1.755 and 2.064 for 10 per cent and 5 per cent selection respectively).

Results and Discussions

Selection in the Base Populations

The results of selection applied in the base populations grown at Bornova and Menemen in 1992-1993 growing season are shown in Table 11.2 and Table 11.3.

Table 11.2: The Estimates of Genetic Gains and the Expected Progeny Population Means Based on the Selection Applied at Bornova in 1992-1993 Growing Season

Population	*Mean (g)*	*Standard Deviation*	*Mean of Selected Group*	*Selection Differential (g)*	*Genetic Gains (g)*	*Expected Plant Yield (g)*
POP-1	1.28	0.67	2.72	1.44	0.75	2.03
AP-1	1.29	0.66	2.68	1.39	0.74	2.03
AP-2	1.29	0.68	2.76	1.44	0.76	2.05
AP-3	1.44	0.77	3.12	1.68	0.86	2.30
POP-K	1.34	0.66	3.02	1.68	0.85	2.19
POP-Q	1.47	0.70	2.89	1.42	0.79	2.26
Kaya	1.41	0.83	3.22	1.81	0.93	2.34
Quantum	1.64	0.87	3.56	1.92	0.98	2.62
Tokak	1.50	0.83	3.24	1.74	0.93	2.43
Efes-1	1.20	0.68	2.70	1.50	0.76	1.96

i = 1.753; H = 0.64

The highest phenotypic standard deviation was 0.87 g for Quantum and the lowest standard deviation was 0.66 for AP-1 mutant population. The highest estimate of genetic gain was 0.98 g for the control Quantum population. The lowest estimate of genetic gain was 0.74 for the AP-1 mutant population. The highest expected means was 2.62 g for quantum. Mutant AP-3 population had 2.3 g expected mean estimate. The control population Efes-1 had the lowest single plant yield (1.96 g).

The expected progeny means were in agreement with the actual mean of the selected groups at Bornova. For example the means of the selected groups were 3.56 g for Quantum, 3.24 g for Tokak and 3.22 g for Kaya populations. AP-1 mutant population had the lowest mean for the selected group.

Table 11.3: The Estimates of Genetic Gains and the Expected Progeny Population Means Based on the Selection Applied at Menemen in 1992-1993 Growing Season

Population	*Mean (g)*	*Standard Deviation*	*Mean of Selected Group*	*Selection Differential (g)*	*Genetic Gains (g)*	*Expected Plant Yield (g)*
POP-1	1.39	0.69	2.82	1.43	0.77	2.16
AP-1	1.68	0.74	3.19	1.51	0.83	2.51
AP-2	1.41	0.59	2.66	1.25	0.66	2.07
AP-3	1.45	0.81	3.19	1.74	0.91	2.36
POP-K	1.27	0.68	2.72	1.45	0.76	2.03
POP-Q	1.45	0.70	2.96	1.51	0.79	2.24
Kaya	1.34	0.58	2.49	1.15	0.65	1.99
Quantum	1.71	0.82	3.35	1.64	0.92	2.63
Tokak	1.68	0.78	3.27	1.59	0.88	2.56
Efes-1	1.44	0.68	2.85	1.41	0.76	2.20

i = 1.753; H = 0.64

Quantum population had highest phenotypic standard deviation 0.82 at Menemen. At this location AP-3 mutant population had high phenotypic standard deviation (0.81). The genetic gains were high for Quantum (0.92) and for AP-3 mutant populations (0.91). The lowest genetic gain was for Kaya. There were some agreements between the expected single plant yields and the actual means of the selected groups at this location. The agreement between two rankings could be accepted as an indicator of suitable selection.

The low single plant yields for the mutant populations were reported by Gaul (1966), Gill *et al.* (1974), Akbay (1977), Yýldýrým *et al.* (1987) and Çaðýrgan (1989). The lower population means than those of controls for single plant yield found in this study were in agreement with these studies.

Selection in the Progeny Populations

The means, range, coefficient of variation and phenotypic standard deviations of the progeny populations grown at Bornova and Menemen in 1993-1994 growing season are shown in Table 11.4.

Table 11.4: The Mean, Range, Coefficient of Variation and Standard Deviations of the Plot Yields for Progeny Populations Grown at Bornova and Menemen in 1993-1994 (n=40 for each population)

Population	Mean (g)		Range		C. V. (per cent)		Phenotypic Std. Dev.	
	Bornova	Menemen	Bornova	Menemen	Bornova	Menemen	Bornova	Menemen
POP-1	84,9±33	67,6±7,0	37.3-149.1	25.9-227.6	24.80	65.60	21.10	44.63
AP-1	84,1±6,5	73,9±6,3	34.3-198.1	23.8-197.0	49.10	53.60	42.40	39.60
AP-2	59,9±4,7	70,4±4,9	23.1-155.8	31.9-161.5	49.30	44.30	29.60	31.20
AP-3	82,3±5,2	103,0±4,6	33.5-186.7	50.2-174.5	39.90	27.90	32.90	28.80
POP-K	55,3±4,6	97,7±5,6	16.6-142.1	35.3-167.3	52.00	36.50	28.80	35.70
POP-Q	54,9±3,7	145,1±5,9	21.4-99.0	53.0-237.5	42.50	25.90	23.40	37.90
Kaya	61,8±3,7	87,9±6,5	13.9-125.6	30.4-223.2	37.70	46.80	23.30	41.30
Quantum	48,3±2,5	138,1±5,1	23.8-91.0	74.2-204.7	32.80	23.30	15.90	32.30
Tokak	59,5±3,30	72,8±4,3	21.5-120.8	40.6-161.2	32.10	37.30	19.10	27.20
Efes-1	43,3±2,8	66,6±6,6	18.1-82.6	20.6-226.4	41.40	62.40	17.90	41.60

It can be seen in Table 11.4 that the mutant populations had higher plot yield means than the control populations. For example, at Bornova POP-1 (84.9 g), AP-1 (84.1 g) and AP-3 (82.3 g) had high means as compared to Kaya (61.8 g), Tokak (59.5 g) and Quantum (72.8 g). Only POP-Q mutant population had higher plot yield than that of the control populations at Menemen (145.8 g vs. 138.1 g for Quantum). It could be concluded that mutant populations had high plot yield as compared with their parents and the control varieties. The range for plot yield was also wider for the mutant populations (34.3 g-198.1 g for AP-1 Vs 18.1 g -82.6 g for Efes-1) at Bornova. The C.V. values indicating high variations were also high for the mutant populations at this location. At Menemen, the control Efes-1 had the widest range (20.6 g-226.4 g) and POP-1 had also wide range for plot yield (25.9 g_227.6 g). The coefficient of variation was the highest for POP-1 (65.6 per cent).

The higher plot yields obtained for mutant populations could be accepted as promising. Therefore, a second stage selection was applied to the mutant population in order to select mutant lines with high yield and agronomical traits as well as protein content. Five percent top yielding progeny rows in each mutant and control population were selected and genetic gain and the expected plot yield were estimated the results of this second stage selections are given in Table 11.5.

The estimates of genetic gains for the mutant progeny populations were higher than those of the control populations. For example, AP-1 (35.0 g), AP-3 (27.8 g), AP-2 (25.1 g), POP-K (24.4 g), POP-Q (19.8 g) mutant populations had higher genetic gains than the control population (the highest estimate for Quantum was 13.5 g). The expected plot yields were also higher for the mutant populations in the same order. POP-1 mutant population had the highest estimate of genetic gain (37.5 g) and the control variety Tokak had the lowest genetic gain (23.0 g). AP-1, POP-K and POP-Q mutant populations had higher genetic gain estimates than Quantum and Tokak control varieties. The mutant progeny population had high expected plot yields such as POP-Q; 177.2 g, POP-K; 127.9 g as compared to Tokak; 95.8 g.

The high genetic gain estimates obtained as a result of the second stage selection applied for plot yield in the mutant populations indicated positive effects of selection on means. The results obtained in this study were in good agreement with earlier reports (Gaul *et al.*, 1969, Atay and Erginel 1976, Moleno-Caro, 1982; Ýýldýrým and Budak, 1990; Ýýldýrým and Çaðýrgan,1991, Çaðýrgan *et al.*, 1998).

Table 11.5: The Estimates of Genetic Gains and the Expected Progeny Plot Yields Based on Selection Applied at Bornova at Menemen in 1993-1994 Growing Season

Population	*Genetic Gain (g)*		*Mean Plot Yield (g)*		*Expected Plot Yield (g)*	
	Bornova	*Menemen*	*Bornova*	*Menemen*	*Bornova*	*Menemen*
POP-1	17.80	37.50	84.90	67.60	102.70	101.10
AP-1	35.00	33.60	84.10	73.90	119.10	107.50
AP-2	25.10	26.40	59.90	70.40	85.30	96.80
AP-3	27.80	24.40	82.30	103.00	110.20	127.40
POP-K	24.40	30.20	55.30	97.70	79.70	127.90
POP Q	19.80	31.40	54.90	145.80	74.70	177.20
Kaya	19.80	34.90	61.90	87.90	81.60	122.90
Quantum	13.50	27.30	48.30	131.10	61.80	165.40
Tokak	16.20	23.00	59.50	72.80	75.70	95.80
Efes-1	15.20	35.20	43.30	66.60	58.50	101.70

i = 2.064; H = 0.41

Agronomical Characteristics of Selected Mutant Lines

The highest top yielding two mutant lines from each population were selected and they were advanced to the micro yield testing in the breeding program. Some agronomical traits of these selections are given in Table 11.6.

It can be seen that in Table 11.6 that certain mutant lines had better agronomic traits as compared to the control varieties. For Example, line TKM-4090, a selection in the POP-Q population, had 237.5 g plot yield, 36 per cent harvest index and 9.9 per cent protein content, TKM-3015, a selection from the POP-1, had high plot yield (227.6 g), high harvest index (31 per cent) and acceptable protein content (9.4 per cent). It can also be seen that selections done in Menemen had better yield and agronomical traits than those of selected at Bornova.

Table 11.6: Means of Certain Traits Pertinent to the Mutant Lines Selected at Bornova and Menemen in 1993-1994 Growing Season

Line Number	*Population Selected*	*Location Selected*	*Plant Height (cm)*	*Spike Length (cm)*	*Kernel Number*	*Plot Yield (g)*	*Harvest Index (%)*	*Protein Content (%)*
TKB-1039	POP-1	Bornova	112.30	14.50	34.00	149.10	0.31	13.70
TKB-2051	POP-1	Bornova	75.80	10.50	30.30	129.50	0.41	10.00
TKM-2099	POP-1	Menemen	93.30	11.30	28.50	182.40	0.24	12.90
TKM-3015	POP-1	Menemen	90.80	9.30	25.00	227.60	0.31	9.40
TKB-1016	AP-1	Bornova	93.30	11.30	33.00	198.10	0.41	12.10
TKB-1090	AP-1	Bornova	86.80	13.50	34.00	185.10	0.42	11.40
TKM-1038	AP-1	Menemen	101.00	10.00	25.00	197.00	0.37	11.50
TKM-2031	AP-1	Menemen	98.30	11.30	31.00	182.60	0.29	11.50
TKB-3069	AP-2	Bornova	81.30	10.80	31.50	139.00	0.32	11.50
TKB-3040	AP-2	Bornova	82.30	10.00	26.50	155.80	0.42	12.40
TKM-1014	AP-2	Menemen	83.80	8.30	22.80	161.50	0.36	8.10
TKM-2074	AP-2	Menemen	95.50	10.30	27.50	147.00	0.32	10.50
TKB-1047	AP-3	Bornova	104.30	11.50	31.30	153.90	0.33	10.70
TKB-1064	AP-3	Bornova	91.00	8.50	24.30	186.70	0.35	12.70
TKM-1028	AP-3	Menemen	96.80	9.80	24.50	174.50	0.35	9.50
TKM-2084	AP-3	Menemen	95.50	10.00	27.50	154.40	0.31	10.50

Contd...

Table 11.6—Contd...

Line Number	Population Selected	Location Selected	Plant Height (cm)	Spike Length (cm)	Kernel Number	Plot Yield (g)	Harvest Index (%)	Protein Content (%)
TKB-1029	POP-K	Bornova	83.50	10.00	27.00	142.10	0.38	12.20
TKB-202	POP-K	Bornova	88.30	11.00	32.00	127.80	0.36	11.20
TKM-206	POP-K	Menemen	96.50	10.50	28.00	166.20	0.29	10.80
TKM-3098	POP-K	Menemen	93.30	8.00	25.00	167.20	0.30	9.50
TKB-1038	POP-Q	Bornova	99.80	10.80	27.50	99.00	0.32	14.10
TKB-4045	POP-Q	Bornova	73.30	10.30	27.80	96.60	0.41	9.90
TKM-3096	POP-Q	Menemen	90.00	11.30	31.50	225.50	0.39	9.60
TKM-4090	POP-Q	Menemen	88.50	11.30	28.50	237.50	0.36	9.90
TKB-2096	Kaya-Control	Bornova	85.00	11.00	30.00	108.70	0.27	12.50
TKM-1067	Kaya	Menemen	107.30	8.30	28.30	212.00	0.41	10.20
TKB-1031	Quantum	Bornova	76.30	9.00	22.30	91.00	0.53	10.10
TKM-2021	Quantum	Menemen	88.80	9.00	25.30	204.70	0.34	9.90
TKM-1975	Efes-3	Menemen	95.00	9.50	26.50	203.20	0.38	10.40
TKB-2031	Tokak	Bornova	88.80	9.80	27.80	120.80	0.29	12.30
TKM-3096	Tokak	Menemen	94.30	10.30	29.80	161.20	0.34	11.50

Selection applied for high individual plant yield in the base population and selection applied for high plot yield in the progeny population resulted in high mutant lines with acceptable agronomical and quality characteristics. Therefore, it may be concluded that mutation breeding in barley will be successful if suitable techniques in constructing the base populations and applying selection in the base and progeny populations are used.

REFERENCES

Allard, R. W. 1960. Principles of Plant Breeding. John Wiley and sons., Inc., New York.

Akbay, G. 1977. EMS (Ethyl Methane Sulphate) uygulanmýþ iki arpa çeþidinin M4 ve M5 hatlarýnda verim ve verim ile ilgili baþlýca karakterier üzerinde araþtýrmalar. Rehabilitation Thesis. Ankara.

Atay, T., Erginel, B. 1976. Gamma iþýnlamasýnýn arpanýn bazý karekterleri üzerindeki etkisi. Bitki, 3:312-317.

Çaðýrgan, M.I. and Yýldýrým, M. B. 1988. Gamma ýþýnlarý uygulanan iki biralýk arpa çeþidinde gözlenen makro mutasyonlar ve bunlardan bitki ýslahýnda yararlanma olanaklarý. IX. Ulusal Biyoloji Kongresi, 21-23. Eylül, 1988. Sivas. Cilt 1. Sf.315-326.

Çaðýrgan, M.I. 1989. Arap mutant populasyonlarýndaki genotipik varyasyonun belirlenmesi ve seleksiyon yoluyla deðerlendirilmesi üzerinde araþtýrmalar. Ph. D. thesis. E. Ü. Fen Bilimleri Enstitüsü. Bornova.

Çaðýrgan, M.I. and Yýldýrým, M. B. 1989. Selection proanthosyanidin-free mutants in a irradiated "Kaya" barley populations. Akdeniz Üniv. Zir. Fak. Dergisi, 2:51-60.

Çaðýrgan, M.I. and Yýldýrým, M. B. 1990. Macro mutational variability in metric traits of barley. Akdeniz Üniv. Zir. Fak. Dergisi, 3:139-152.

Çaðýrgan, M.I. K., Visser, Toker, C., Wang, M., Yýldýrým, M.B., Tugay, M. E. and Heidekamp, F. 1998. Registration of M-Q-54 a fast germinating ABA-insensitive of mutant of "Quantum" barley. Crop Sci, 38:1410.

Efe, H. and Yýldýrým, M.B. 1992. Bazý mutant arpa hatlarýnda verim ve verim componentleri üzerinde bir araþtýrma. II. Barley-Malt Seminar. Konya. P.265-270.

Garcia del Moral, L.F., Ramos, J.M., Graica del Moral, M.B. and Jimenez-Fejada, M. P. 1991. Ontogenic approach to grain production in spring barley based on path coefficient analysis. Crop Sci., 31:1179-1185.

Gaul, H. 1963. Mutationen in der Pflanzenzüchtung. 1 and 2 Teil. Z. Pflazenzüchtung, 50:194-246.

Gaul, H. 1966. Züchterische Bedeutung von Klienmutationen. I. Durch Röntgenstrahlen induziarte Variabilitat von Kornertrag Korngrose und Vegetationslange bei der Gerste Haisa 11, Z. Pflanzenzüchtug. 55:1-20.

Gaul, H., Ulonska, E., Winkel, C. and Braker, G. 1969. Micro mutations influencing yield in barley, studies over nine generations. Induced Mutations in Plants. IAEA. Vienna. pp.375-398.

Gill, K. S., Bains, K. S. and Chand, K. 1974. Differential response of mutagenes in inducing mutation genetic variation in metrical traits in barley. Z. Pflanzenzüchtug, 71:117-123.

Molina-Cano, J.L. 1982. Genetic, agronomic and malting characteristics of a new erectoide mutants induced in barley. Z. Pflanzenzüchtug, 88:34-42.

Steel, R.G.D. and Torrie, J. H. 1980. Principles and Procedures of Statistics. Second Edition. Mc Graw-hill Book Comp. Inc. N.Y.

Thachuk, R. 1981. Protein analysis of whole wheat kernels by Near Infrared Reflectance. Cereal Food World. 26: 584-587.

Ýbrahim, A. F. and Shaaran, A. N. 1974. Variability of character expression in barley. M3 and M4 bulk populations after seed irradiation with gamma rays. Z. Pflanzenzüchtug, 72:212-225.

Yýldýrým, M. B., Çaðýrgan, M.I. and Turgut, I. 1987. Arpa mutant populasyonlarýnda seleksiyon uygulamasý. Türkye Tahýl Simpozyumu. Bursa. pp.473-487.

Yýldýrým, M. B. and Budak, N. 1990. Kaya arpa çeþidinin mutant populasyonlarýndan seçilmiþ hatlarýn döl kontrolu sonuçlarý. E. U. Zir. Fak. Dergisi, 26:79-86.

Yýldýrým, M. B. and Çaðýrgan, M.I. 1991. Selection studies for grain yield in certain mutant populations of barley. Akdeniz Üniv. Zir. Fak.. Dergisi, 4:1-10.

Yýldýrým, M. B., Budak, N. and Tugay, E. 1996. A study on the yield and protein contents of mutant barley lines. Turkish J. Field Crops., 1:24-32.

Chapter 12

Linkage Mapping Using Mutant Genes in Rice

T. Kinoshita

Faculty of Agriculture, Hokkaido University, Sapporo, 060-0809, Japan

ABSTRACT

Rice linkage maps using conventional markers were constructed by using a wealth of genetic data published hitherto. By using available markers over 900, 552 of them were assigned to the 12 linkage groups corresponding to the haploid chromosome number.

Among them, mutant characters occurring from spontaneous or induced mutations have contributed to the linkage analysis for adequate markers. It is noted that most of the spontaneous mutations have been induced with various physical and chemical mutagens. Recently a large-scale sequence analysis of rice cDNAs from callus and various organs generated a number of ESTs (non-redundant expressed sequence tags) to identify biochemical marker genes.

Further, high resolution linkage maps were constructed independently for a relatively short term in several groups. Therefore all efforts are needed to integrate all existing markers into one unified, densely populated map. Along this line, conventional markers were tagged with molecular markers such as RFLP, RAPD, AFLP and microsatellite markers. In economically important characters, tagging genes for disease and insect resistances led to marker-aided selection and marker-based gene cloning.

Mapping of quantitative trait loci (QTLs) and construction of physical maps with YAC and BAC clones are also important resources for positional cloning and genome analysis in detail. In situ hybridization is another effective means for localizing specific DNA on chromosome.

Because of its having the smallest genome among cereals and the development of transformation techniques, rice is earmarked as a model crop for research in molecular genetics. Many promising technologies for rice breeding will be developed on the basis of rice genome analysis.

Because rice is an important food for nearly half the world's population, a considerable amount of data concerning rice genetics and breeding has been accumulated by many workers (Khush and Kinoshita 1991, Kinoshita 1984, Matsuo *et al.*, 1997).

Construction of linkage maps has also progressed steadly. However, owing to confusion due to inadequate knowledge of genetical analysis of complex characters such as anthocyanin coloration, linkage studies were delayed for a considerable time in comparison with other major food crops such as maize and barley (Nagao 1951).

By the efforts of rice geneticists in the world, the number of marker genes has reached over 900 and 552 of them have been assigned to the 12 chromoseomes corresponding to the haploid chromosome number of rice.

Marker Genes

Depending on the gene analysis of various morphological, physiological and biochemical characters, the numbers of marker genes classified into 31 genetic traits are shown in Table 12.1. Many of these were identified as naturally occurring variations of spontaneous origin in varietal populations; others were induced through mutagenic treatments. The morphological markers affecting various plant parts such as roots, stems, leaves, panicles, florets and kernels are easiest to identify and have been thoroughly investigated. Genes for disease and insect resistance can be identified

Table 12.1: Classification of Genetic Traits and the Numbers of Marker Genes

Class No.	*Name*	*Number of Marker Genes**
1.	Anthocyanin coloration	25
2.	Inhibitor for anthocyanin coloration	11
3.	Coloration other than anthocyanin	16
4.	Chlorophyll aberration	67
5.	Small grain dwarf	8
6.	Tillering dwarf (High-tillering dwarf)	8
7.	Malformed dwarf	19
8.	Semidwarf	20
9.	Awn	8
10.	Lemma and palea	17
11.	Spikelet and grain	34
12.	Endosperm	69
13.	Panicle	27
14.	Mode of branching	4
15.	Culm	35
16.	Leaf	31
17.	Leaf spot	19
18.	Root	7
19.	Heading date	31
20.	Sterility	104
21.	Hybrid weakness	14
22.	Gametophyte gene	15
23.	Cytoplasmic male sterility (Cytoplasm)	0 (19)
24.	Fertility restorer	10
25.	Fungal and bacterial disease resistance	77
26.	Virus and mycoplasma disease resistance	10
27.	Insect resistance	40
28.	Isozyme	67
29.	Abiotic stress tolerance	13
30.	Biochemical characters (cloned genes)	69**
31	Embryo	26
	Total	**901**

* Most of the marker genes are maintained in Japan and IRRI.

** Most of the genes analyzed from randomly chosen cDNA (ESTs) are not included.

only on the basis of tests using pathogens. Recently biochemical and phyiological genes are increasing rapidly. Besides the usual gene analyses, a number of cloned genes and cDNAs from callus, endosperm, roots and other tissues were identified in a large-scale sequence analysis (Kurata *et al.*, 1994, Liu 1995, Sasaki 1996, Yamamoto and Sasaki 1997).

On account of new developments in rice genetics, the need for examining gene symbols and linkage groups has been progressively intensified. For this reason, Nagao (Nagao 1935, 1951) and Kadam and Ramiah (1943). proposed a unified system of gene symbols.

Following the international rule for gene symbolization of all organisms, which was accepted by the Tenth International Genetics Congress, the working group of the International Rice Commission (IRC) provided guidelines for gene symbolization and a list of suggested gene symbols (Crops Research 1963). Finally, the Rice Genetics Cooperative (RGC), which was established to promote cooperation in rice genetics in 1985, adopted new rules for gene symbolization (Kinoshita 1986). Since then, one of the standing committees of RGC coordinates and monitors gene symbols. New gene symbols and revised linkage maps are published annually in Rice Gemetics News Letter (RGN).

Table 12.2 lists the marker gens alphabetically along with linkage groups (chromosomes), genetic nature and mutagenesis except for the genes not allotted to the linkage groups. It is noted that many mutant genes have an important role as genetic markers. Several genes such as d1 (daikoku dwarf), d7 (cleistogamous dwarf), d18 (Hosetsu dwarf), sd1 (Dee-geo-woo-gen dwarf), lg (liguleless) and la (lazy growth habit) repeatedly occurred by spontaneous and induced mutations. Physical and chemical mutagens were extensively used in various materials and ecomomically important mutants were also induced in semidwarfing, heading date, grain quality and male sterility.

In spite of efforts of coordinators in RGC, there are still some problems on gene symbolisations. In addition there are many unauthorized genes which were marked with asterisks in the table. Above all prompt registration with the convener is expected in accordance with rule 7 (Kinoshita 1986). Thus the confusion related to the numbering and name of genes will be avoided.

Table 12.2: List of Marker Genes Assigned to the Linkage Groups

Gene Symbol	Trait No.	Chromosome No.	Name/Alelles/Gene Interaction/Mutagenesis[1]
A(Sp)	1	1	Anthocyanin activator/*A–S, A–E, A–d. A–m*/complementary with *C* and *P*
*ACC1**	30	3	ACC synthase–1
*ACC3**	30	5	ACC synthase–3
Acp1	28	12	Acid phosphatase–1/*Acp1–0, Acp1–1(Acp1–4), Acp1–2 (Acp1––+9), Acp1–3(Acp1––+4)*
Acp2	28	12	Acid phosphatase–2/*Acp2–0, Acp2–1(Acp2–Fa)*
Acp4	28	7	Acid phosphatase–4/*Acp4-1, Acp4-2*
Actin1(act1)*	30	3	Actin–1
Adh1	28	11	Alcohol dehydrogenase–1/*Adh1–0, Adh1–1, Adh1–2, Adh1–3*
*Adh2**	28	9	Alcohol dehydrogenase–2
*adh2(t)**	30	11	alcohol dehydrogenase–2(cDNA)
*ae3(t)**	12	2	amylose extender–3/MNU[1]
al1(alK1)	4	6	albino–1/MNU
al2	4	5	albino–2/MNU
al3	4	5	albino–3/MNU
al4	4	1	albino–4/MNU
al5	4	4	albino–5/MNU
al6(t)	4	5	albino–6/MNU
al7(t)	4	4	albino–7/MNU
al8	4	1	albino–8/MNU
al9(t)	4	6	albino–9/MNU
al10	4	3	albino–10/MNU
*Alu1**	30	5	Aldolase–1
Ald2(ald2)*	30	1	Aldolase–2
Ald3(alu1)*	30	12	Aldolase–3
alk	12	6	alkali degeneration
Amp1(Lap1, LapF)	28	2	Aminopeptidase–1/*Amp1–0, Amp1–1, Amp1–2, Amp1–3, Amp1–4, Amp1–5, Amp1–6*
Amp2(AlapA)	28	8	Aminopeptidase–2/*Amp2–1, Amp2–2, Amp2–3, Amp2–4*

Table 12.2–Contd...

Gene Symbol	*Trait No.*	*Chromo-some No.*	*Name/Alelles/Gene Interaction/Mutagenesis'*
Amp3(Amp1)	28	6	Aminopeptidase–3/*Amp3–0, Amp3–1, Amp3–2, Amp3–3, Amp3–4, Amp3–5, Amp3–6*
Amp–4	28	8	Aminopeptidase–4/*Amp4–1, Amp4–2, Amp4–3, Amp4–4*
Amp–5	28	6	Aminopeptidase–5/*Amp5–0, Amp5–1, Amp5–2, Amp5–3, Amp5–4*
Amy1A/C (RAmy1A/C)*	30	2	Alpha–amylase–1A/C
Amy1B(RAmy1B)*	30	1	Alpha–amylase–1B
Amy2A(RAmy2A)*	30	6	Alpha–amylase–2A
Amy3A/B/C (RAmy3A/B/C)*	30	9	Alpha–amylase–3A/B/C
Amy3D/E (RAmy3D/E)*	30	8	Alpha–amylase–3D/E
An1	9	4	Awn–1/duplicate or triplicate
An2	9	5	Awn–2
An3	9	3	Awn–3
*ant**	30	2	adenine nucleotide trarslocator
aph	10	6	apiculus hairs
*Arf1**	30	5	ADP ribosylation factor (ARF) –1
*Arf2**	30	1	ADP ribosylation factor (ARF) –2
ATP(atp1)*	30	5	ATPase
*atub**	30	9	Alpha tubulin
aul	16	4	auricleless/spontaneous
bc1	15	3	brittle culm–1/spontaneous, MNU
bc2	15	5	brittle culm–2/internal irradiation with P^{32}
bc3	15	2	brittle culm–3/gamma ray
bc4	15	6	brittle culm–4/unknown
bc5	15	2	brittle node/unknown
bd1	10	5	beaked lemma–1/duplicate/unknown
betaAmy1	28	7	Beta Amylase isozyme–1/*BAmy1–0, BAmy1–1, BAmy1–2, BAmy1–3*
bgl	4	5	bright green leaf/MNU

Contd...

Table 12.2–Contd...

Gene Symbol	Trait No.	Chromo-some No.	Name/Alelles/Gene Interaction/Mutagenesis[1]
bl1	17	2	brown leaf spot–1 /MNU
bl2(blm)	17	6	brown leaf spot–2/spontaneous
bl3	17	6	brown leaf spot–3/spontaneous
bl4	17	3	brown leaf spot–4/internal irradiation with P^{32}
Bp	13	9	Burlush-like panicle/spontaneous
Bph1	27	12	Brown planthopper resistance–1
bph2	27	12	brown planthopper resistance–2
Bph3	27	4	Brown planthopper resistance–3
bph4	27	4	Brown planthopper resistance–4
Bph9	27	12	Brown planthopper resistance–9
*Bph10(t)**	27	12	Brown planthopper resistance introduced from *Oryza australiensis*
*Bph(t)**	27	9	Brown planthopper resistance
C	1	6	Chromogen for anthocyanin/*C–Bs, C–B, C–Bp, C–Bt, C–Br, C–Bd, C–Bk, C–Bc, C–Bm*/complementary with A and p
*CALa**	30	5	Calmodulin–a
*CALb**	30	7	Calmodulin–b
*CatA1**	30	3	Catalase–AI (cDNA clone)
Cat1	28	6	Catalase–1/*Cat1–0, Cat1–1, Cat1–2, Cat1–3*
*cdc**	30	3	CDC2a protein
ch1(ch1)	4	3	chlorina–1/spontaneous
chl2	4	3	chlorina–2/gamma ray
chl3	4	3	chlorina–3/spontaneous
chl4	4	6	chlorina–4/gamma ray
chl5	4	1	chlorina–5/MNU
chl6	4	1	chlorine–6/MNU
chl7	4	6	chlorina–7/induced
chl8	4	8	chlorina–8/unknown
chl9	4	8	chlorina–9/unknown

Contd

Table 12.2–Contd...

Gene Symbol	Trait No.	Chromo-some No.	Name/Alelles/Gene Interaction/Mutagenesis[1]
chl10	4	2	chlorina-10/MNU
*chs**	30	11	chalcone synthase
*chs1(t)**	4	9	chlorosis caused by low temperature–1(17ºC)/*chs1–A, chs1–B, chs1–C*
Cht1(Chi1)*	30	6	Chitinase–1
*Cht2**	30	5	Chitinase–2
*Cht3**	30	6	Chitinase–3
Cht4(Chi4)*	30	4	Chitinase–4
Cinf(t)(Lcr)*	20	6	Cross–incompatibility in the female reaction
Cl	14	6	Clustered spikelets/spontaneous
*cyc1**	30	9	cyclophilin–1
*cyc2**	30	2	cyclophilin–2
*D1**	21	2	Dominant lethal from *O. longistaminata*/complementary
*D2**	21	11	Dominant lethal from *O. sativa*
d1	5	5	daikoku dwarf/spontaneous, gamma ray, MNU
d2	7	1	ebisu dwarf/spontaneous
d3	6	4	bungetsuwaito tillering dwarf/duplicate or triplicate/spontaneous
d4	6	6	bungetsuwaito tillering dwarf/spontaneous
d5	6	2	bungetsuwaito tillering dwarf/spontaneous
d6(d34)	7	7	ebisumochi dwarf or tankanshirazasa dwarf/spontaneous
d7	5	7	heieidaikoku or cleistogamous dwarf/spontaneous, gamma ray, EI, MNU
d9	8	6	Chinese dwarf/spontaneous
d10(d15, d16)	6	1	kikeibanshinriki or toyohikaribunwai tillering dwarf/spontaneous
d11(d8)	5	4	shinkaneaikoku or nohrin 28 dwarf/spontaneous, internal irradiation with P^{32}, EI

Contd...

Table 12.2–Contd...

Gene Symbol	Trait No.	Chromo-some No.	Name/Alelles/Gene Interaction/Mutagenesis[1]
d14(d10)	6	3	kamikawabunwai tillering dwarf/ spontaneous
d18(d25)	7	1	hosetsuwaisei or akibare dwarf and kotaketama–nishiki dwarf/*d18–h, d18–k*/ spontaneous, gamma ray, EI
d20	7	3	hayayuki dwarf/spontaneous, gamma ray
d21	7	6	aomorimochi–14 dwarf/spontaneous
d26	7	1	7237 dwarf/unknown
d27(d t)	6	11	bungetsuto tillering dwarf/spontaneous
d28(d C)	5	11	chokeidaikoku or long stemmed dwarf/ spontaneous
d29(d K1)	7	2	short uppermost internode dwarf/gamma ray
d30(d W)	5	2	waiseishirazasa dwarf/spontaneous
d31	7	4	taichung–155–irradiated dwarf/X ray
D32(d K4, d12)	7	2	dwarf Kyushu–4/gamma ray
d33(d B)	6	12	bonsaito dwarf/spontaneous
d42	7	4	liguleless dwarf/X ray, gamma ray
d51(d K8)	7	8	dwarf Kyushu–8/MNU
d52(D K2)	7	3	dwarf Kyushu–2/MNU
D53(D K3)	7	11	dwarf Kyushu–3/gamma ray
d54(d K5)	7	1	dwarf Kyushu–5/MNU
d55(d K6)	7	1	dwarf Kyushu–6/MNU
d56(d K7)	7	3	dwarf Kyushu–7/MNU
d57[d(x)]	7	9	dwarf/unknown
d58	5	6	small grained dwarf/gamma ray
d60[sd(t)]	8	7	dwarf (Hokuriku 100)/gamma ray
dl(lop)	16	3	drooping leaf/*dl, dl–1, dl–2, dl–sup1, dl–supp.*/spontaneous, MNU
Dn1	13	9	Dense panicle–1/spontaneous
dp1	10	6	depressed palea–1/spontaneous
dp2	10	0	depressed palea–2/spontaneous, gamma ray

Contd...

Table 12.2–Contd...

Gene Symbol	Trait No.	Chromo-some No.	Name/Alelles/Gene Interaction/Mutagenesis[1]
*dp(t)**	10	9	depressed palea/spontaneous
drp1	16	4	dripping-wet leaf–1/spontaneous
drp2	16	9	dripping-wet leaf–2/MNU
drp3	16	3	dripping-wet leaf–3/MNU
drp4	16	3	dripping-wet leaf–4/MNU
drp5	16	4	dripping-wet leaf–5/MNU
drp6	16	6	dripping-wet leaf–6/MNU
drp7	16	11	dripping-wet leaf–7/MNU
drp8	16	4	dripping-wet leaf–8/MNU
ds3(t)	20	7	desynapsis–3/MNU
*Dse**	18	11	Diameter of seminal root
du1	12	10	dull endosperm–1/MNU
du4	12	12	dull endosperm–4/MNU
*du2035**	12	6	dull endosperm (2035)/EMS
*du2120**	12	9	dull endosperm (2120)/EMS
*duEM47**	12	6	dull endosperm (EM47)/EMS
dw1(fh1)	15	6	deep water tolerance–1/duplicate
El(=m Ef1+)	19	7	Heading date–1/modifier for *Ef1(mEf1–a, mEf1–b)*
E3	19	3	Heading date–3
Ef1(Ef2)	19	10	Earliness–1/*Ef1, Ef1–a, Ef1–b, Efl–gamma, Ef1–x, Ef1–n*/X ray, gamma ray
*Efx(t)**	19	2	Earliness-x/complementary
*Efy(t)**	19	6 or 7	Earliness-y
*Ef(t)**	19	10	Earliness introduced from *Oryza australiensis*
eg	10	1	extra glumel spontaneous, MNU
*elf2**	30	7	elongation factor 1 beta chain–2
*Em**	30	5	ABA–regulated gene/expressed late in embryogenesis
EnSe1(t)	19	6	Enhancer for photoperiod-sensitivity (*Se1*)

Contd...

Table 12.2–Contd...

Gene Symbol	Trait No.	Chromo-some No.	Name/Alelles/Gene Interaction/Mutagenesis[1]
*Enp1**	28	6	Endopeptidase–1/*Enp1–0, Enp1–1*/ monomer
er(o)	15	5	erect growth habit
esp1(rsp1)	12	7	endosperm storage protein–1/MNU
esp2(rsp2)	12	11	endosperm storage protein–2/MNU
esp3(rsp3)	12	11	endosperm storage protein–3/MNU
Est1	28	7	Esterase–1/*Est1–0, Est1–1(Est1–S)*
Est2	28	6	Esterase–2/*Est2–0, Est2–1(East2–S), Est2–2(Est2–F)*
Est3	28	9	Esterase–3/*Est3–1(Est3–S), Est3–2(Est3–F)*
Est5	28	1	Esterase–5/*Est5–0, Est5–1, Est5–2*
Est7	28	7	Esterase–7/*Est7-0, Est7-1*
Est9(Estcl)	28	7	Esterase–9/*Est9–0, Est9–1, Est9–2*
Est10	28	1	Esterase–10/ *Est10-0, Est10–1, Est10–2, Est10-3, Est10-4, Est10-5*
Est12	28	9	Esterase–12/*Est12–0, Est12–1*
Est13	28	5	Esterase–13/*Est13–1, Est13–2*
*Estl2**	28	1	Esterase–12/*Estl2–aa, Estl2–bb, Estl2–nn*
eui	15	5	elongated uppermost internode/ spontaneous
fc1	15	3	fine culm–1/gamma ray
fc2	15	6	fine culm–2/MNU
*Fdp1**	28	11	Fructose–1, 6–diphosphatase1/*Fdp1–0, Fdp1–1, Fdep1–2*
fgl(fl)	4	10	faded green leaf/spontaneous, MNU
flo1	2	5	floury endosperm–1/MNU
flo2	12	4	floury endosperm–2/gamma ray, MNU
fs1	4	6	fine stripe–1/X ray, gamma ray
fs2	4	1	fine stripe–2/internal irradiation with P^{32}
g1(lng)	10	7	long sterile lemmas–1/spontaneous, MNU
ga1	22	6	gametophyte gene–1/Atomic bombed

Contd...

Table 12.2–Contd...

Gene Symbol	Trait No.	Chromo-some No.	Name/Alelles/Gene Interaction/Mutagenesis[1]
ga2	22	3	gametophyte gene–2/*Ga2–A, Ga2–B, Ga2–C*
ga3	22	3	gametophyte gene–3
ga4(gaA)	22	6	gametophyte gene–4/gamma ray
ga5(gaB)	22	6	gametophyte gene–5
ga6	22	4	gametophyte gene–6/gamma ray
ga7	22	1	gametophyte gene–7
ga8	22	1	gamotophyte gene–8
ga9	22	1	gametophyte gene–9
ga10(t)	22	4	gametophyte gene–10/EI
ga11	22	7	gametophyte gene–11/*ga14–i, ga11–a*
ga12	22	4	gametophyte gene–12/MNU
ga13	22	12	gametophyte gene–13
ga14	22	3	gametophyte gene–14/*ga14–i, ga14–j, gal14–n*
*gax**	22	8	gametophyte gene–x
gal(lt)	20	11	gametio lethal
*gcw2**	30	10	glycin-rich cell wall structural protein–2
*gcw4**	30	10	glycin-rich cell wall structural protein–4
Gdh1	28	3	Glutamate dehydrogenase–1/*Gdh1–1, Gdh1–2*
ge1	31	7	giant embryo–1/*ge1, ge1–1, ge1–2, ge1–3, ge1–4, ge1–5, ge1–6, ge1–s/MNU*
gf1	3	6	gold furrows of hull–1
gf2	3	1	gold furrows of hull–2/spontaneous
gh1	3	5	gold hull and internode–1/spontaneous
gh2	3	2	gold hull and internode–2/MNU
gh3	3	2	gold hull and internode–3/gamma ray
gl1	16	5	glabrous leaf and hull–1/duplicate/ spontaneous, somaclonal, mutant with anther culture
Glh3	27	4	Green leafhopper resistance–3
Glh6	27	5	Green leafhopper resistance–6

Contd...

Table 12.2–Contd...

Gene Symbol	Trait No.	Chromo-some No.	Name/Alelles/Gene Interaction/Mutagenesis[1]
Glh(t)*	27	4	Green leafhopper resistance
gku1(t)*(Lgc1)	12	2	lack of glutelin subunit–1
glu2(t)*	12	10	lack of glutelin subunit–2
glu3(t)*	12	1	lack of glutelin subunit–3
glu4(t)*	12	1	lack of glutelin subunit(alpha–2)-4/glu4–a, glu4–b, glu4–c
Glu1(Srp1, Gt2)	12	1	Rice glutelin–1/Glu–1, Glu1–2
Glu1t*(Glu1)	12	1	Glutelin subunit–1
Glu2t*(Glu2)	12	1	Glutelin subunit–2
Glu3t*(Glu3)	12	1	Glutelin subunit–3
Glu4t*(Glu4)	12	2	Glutelin subunit–4
Glu5t*(Glu5)	12	2	Glutelin subunit–5
Glu6t*(Glu6, GluB, C, E)	12	2	Glutelin subunit–6
Glu7t*(Glu7, GluA)	12	2	Glutelin subunit–7
Glup1(Esp5, Gup1)	12	9	Glutelin precursor–1/MNU
glup2(esp6, gup2)	12	9	glutelin precursor–2/MNU
glup3(esp7)	12	4	glutelin precursor–3/spontaneous
glup4(esp8)	12	5	glutelin precursor–4/MNU
Glup5(t)*	12	9	Glutelin precursor–5/MNU
Glup6(t)*	12	6	Glutelin precursor–6/MNU
Gm2	27	4	Gall midge resistance–2
gm1, gm2(pd)	27	9	gall midge resistance/duplicate
Got1	28	1	Aspartate aminotransferese–1/Got1–1, Got1–2, Got1–3
Got2	28	6	Aspartate aminotransferase–2/Got2–1, Got2–2
Got3	28	2	Aspartate aminotransferese–3/Got3–1, Got3–2
gpd1*	30	2	glyceraldehyde–3–phosphate dehydrogenase–1
gpd2*	30	4	glyceraldohyde–3–phosphate dehydrogenase–2

Contd...

Table 12.2–Contd...

Gene Symbol	Trait No.	Chromo-some No.	Name/Alelles/Gene Interaction/Mutagenesis[1]
Grh1	27	5	Green rice leafhopper resistance–1/ duplicate or complementary
Grh2(Grha)	27	11	Green rice leafhopper resistance–2
*Grh3(t)**	27	6	Green rice leafhopper resistance–3
Grh4(t)(Grhb)*	27	3	Green rice leafhopper resistance–4/ complementary with *Grh2*
Hbv	26	12	Hoja blanca resistance
*Hd2(t)**	19	7	Heading date–2
*Hd3(t)**	19	6	Heading date–3
*Hd4(t)**	19	7	Heading date–4
*Hd5(t)**	19	8	Heading date–5
*Hd6(t)**	19	3	Heading date–6
Hg	10	3	Hairy glume/spontaneous
Hla	16	6	Hairy leaf–a/complementary spontaneous
*hpg1**	12	4	high preproglutelin–1
*hsp2**	30	9	heat shock protein–2
*hsp4**	30	9	heat shock protein–4
hwd1	21	10	hybrid weakness–d/complementary
hwd2	21	7	hybrid weakness–d
IBf	3	9	Inhibitor for brown furrows/spontaneous, MNU
Icd1	28	1	Isocitrate dehydrogenase–1/*Icd1–1, Icd1–2, Icd1–3(IcdA)*
*IPi(t)**	25	12	Pyricularia oryzae resistance
*Ipi3(t)**	25	12	Pyricularia oryzae resistance
IPl1	2	5	Inhibitor for purple leaf–1/duplicate or triplicate
IPl2	2	6	Inhibitor for purple leaf–2
IPl4	2	6	Inhibitor for purple pericarp/complementary
IPsb(IPs2)	2	1	Inhibitor for purple stigma–b
*L**	20	3	Lethal factor

Contd...

Table 12.2–Contd...

Gene Symbol	Trait No.	Chromo-some No.	Name/Alelles/Gene Interaction/Mutagenesis[1]
La	15	11	lazy growth habit/spontaneous, gamma ray, MNU
lam(t)	12	9	low amylose endosperm/gamma ray + EMS
lax(lx)	13	1	lax panicle/spontaneous
*Ldh**	30	6	Lactate dehydrogenase
*Lec**	30	12	Lectin
lg	16	4	liguleless/*lg–a*/spontaneous, gamma ray, internal irradiation with P^{32}, EMS, MNU
lgt	11	1	long twisted grain/unknown
*lh(t)**	19	7	late heading
lhs(lhs2,op)	20	3	leafy hull sterile/spontaneous, MNU
Lk2(t)(l knb)	11	11	'Nagayama 77402b (N 182)' long grain/ spontaneous
Lkf	11	3	'Fusayoshi' long grain/spontaneous, gamma ray + EMS
lki1	11	4	'IRAT 13' long grain–1/duplicate/ spontaneous
lp1	10	7	long palea–1/duplicate/spontaneous
Mal1	28	7	Malate dehydrogenase (NADP)1/*Mal1–1, Mal1–2*
Mal2	28	7	Malate dehydrogenase (NADP)2/*Mal2–1, Mal2–2*
Man	12	9	Mannan in endosperm cell wall
*Mdh3**	28	5	Malate dehydrogenase (NAD)–3
Mi	11	3	Minute grain/spontaneous
mik	11	11	'Kitaake mutant' minute grain/gamma ray
mis3	10	4	malformed spikelet–3/unknown
mp1	11	1	multiple pistil–1/unknown
mp2	11	6	multiple pistil–2/unknown
MPiz(Rb6)	25	11	Modifier for blast resistance (*Piz*)
MPox1(MPx1)	28	5	Modifier for peroxidase *(Pox1)/MPox1–0, MPox1–1*
ms1(sf)	20	6	male sterile–1/spontaneous

Contd...

Table 12.2–Contd...

Gene Symbol	Trait No.	Chromo- some No.	Name/Alelles/Gene Interaction/Mutagenesis[1]
ms7	20	3	male sterile–7/EI
ms8	20	7	male sterile–8/EI
ms9	20	6	male sterile–9/EI
ms10	20	9	male sterile–10/EI
ms14	20	5	male sterile–14/gamma ray
ms17	20	2	male sterile–17/gamma ray
msm67(t)(ms18)*	20	1	male sterile (Milyang 67ms)/spontaneous
*msn77(t)**	20	1	male sterile (Milyang 77ms)/spontaneous
nal1	16	4	narrow leaf–1/duplicate or triplicate/ spontaneous
nal2	16	11	narrow leaf–2/spontaneous
nal3(nal2)	16	12	narrow leaf–3/spontaneous
nal4(nal)	16	4	narrow leaf–4/spontaneous
nal5(nal1)	16	4	narrow leaf–5/gamma ray
Nal6	16	3	narrow leaf–6/MNU
nl1	13	5	neck leaf–1/spontaneous
nl2	13	5	neck leaf–2/gamma ray
*Oc1**	30	1	Oryzacystatin–1
*Oc2**	30	5	Oryzacystatin–2
*ocp**	30	4	oryzain alpha chain
ops	20	5	open hull sterile/spontaneous
*OsFAD3**	30	11	w-3 fatty acid desaturase–3
*OsFADX**	30	12	w-3 fatty acid desaturase–X
*OsFADY**	30	11	w-3 fatty acid desaturase–Y
P(Pa, Pb, Pc)	1	4	Colored apiculus/*P–K*, *P–Cl* complementary with *C* and *A*
*Pal1**	30	2	Phenylalanine ammonia-lyase
Pau	1	1	Purple auricle
*PBR**	25	11	Panicle blast resistance
*Pbst**	25	11	Panicle blast resistance
Pgd1	28	11	Phosphogluconate dehydrogenase–1/ *Pgd1–1, Pgd1–2, Pgd1–3*

Contd...

Table 12.2–Contd...

Gene Symbol	Trait No.	Chromo-some No.	Name/Alelles/Gene Interaction/Mutagenesis[1]
Pgd2	28	6	Phosphogluconate dehydrogenase–2/ *Pgd2–1, Pgd2–2*
Pgi1(Pgia)	28	3	Phosphoglucoisomerase–1/*Pgi1–0, Pgi1–1, Pgi1–2, Pgi1–3*
Pgi2(Pgib)	28	6	Phosphoglucoisomerase–2/*Pgi2–1, Pgi2–2, Pgi2–3, Pgi2–4*
pgl	4	10	pale green leaf/spontaneous, MNU
Ph(=Bhc, Po)	3	4	Phenol staining
*PHYA**	30	3	Phytochrome–A
Pia	25	11	Pyricularia oryzae resistance–a
Pib(pis)	25	2	Pyricularja oryzae resistance–b
Pif	25	11	Pyricularia oryzae resistance–f
Plls1(Rb4)	25	11	Pyricularia oryzae resistance–is1/ cumulative
Pik	25	11	Pyricularia oryzae resistance–k/*Pik, Pik–s, Pik–p, Pik–m, (Pim), Pik–h, Pik–g(t)*
*Pikur1**	25	4	Pyricularia oryzae resistance–kur1/field resistance/additive
*Pikur2**	25	11	Pyricularia oryzae resistance–kur2
Pise1(Rb1)	25	11	Pyricularia oryzae resistance–se1/ additive
Pit	25	1	Pyricularia oryzae resistance–t
Pita(sl, Pi4a)	25	12	Pyricularia oryzae resistance–ta/*Pita, Pita–2, Pita–n*
Piz(Pi2)	25	6	Pyricularia oryzae resistance–z/*Piz, Piz–t, Piz–5*
*Pi(t)**	25	4	Pyricularia oryzae resistance
Pi1	25	11	Pyricularia oryzae resistance–1
Pi5(t)		4 or 9	Pyricularia oryzae resistance–5
*Pi6(t)**	25	12	Pyricularia oryzae resistance–6
Pi7(t)	25	11	Pyricularia oryzae resistance–7
Pi8	25	6	Pyricularia oryzae resistance–8
*Pi9(t)**	25	6	Pyricularia oryzae resistance–9
*Pi10(t)**	25	5	Pyricularia oryzae resistance–10

Contd...

Table 12.2–Contd...

Gene Symbol	Trait No.	Chromo-some No.	Name/Alelles/Gene Interaction/Mutagenesis[1]
Pi11(t)(Pizh)*	25	8	Pyricularia oryzae resistance–11
*Pi12**	25	12	Pyricularia oryzae resistance–12 from Hong-jiao zhan
Pi13(t)	25	6	Pyricularia oryzae resistance–13
*Pi13**	25	6	Pyricularia oryzae resistance–13
Pi14(t)	25	2	Pyricularia oryzae resistance–14
*Pi14**	25	12	Pyricularia oryzae resistance–14
Pi16(t)	25	2	Pyricularia oryzae resistance–16
Pi17(t)	25	7	Pyrioularia oryzae resistance–17
*Pi18(t)**	25	11	Pyricularia oryzae resistance–18
Pi62(t)(Pita?)*	25	12	Pyricularia oryzae resistance from Yashiro-mochi
*Pi157**	25	12	Pyricularia oryzee resistance from Morobereken/complete resistance to an Indica isolate, B157 of the blast fungus
Pin1	1	4	Purple internode–1
*pkc1**	30	9	protein kinase C homology–1
*pkc2**	30	8	protein kinase C homology–2
P1	1	4	Purple leaf/*Pl, Pl–w, Pl–9(Pl')*
pms1 (Pms1, pgms, ms1P)*	20	7	photoperiod-sensitjve male sterility–1/ sponteneous
pms2 (Pms2, pgms, ms2P)*	20	3	photoperiod-sensitive male sterility–2/ spontaneous
*pms**	20	6	photoperiod-sensitive male sterility/ unknown
Pn	1	1	Purple node
Pox1(PxPe)	28	5	Peroxidase–1/*Pox1–0, Pox1–1(Pox1–2A), Pox1–2(Pox1–0C)*
Pox2	28	12	Peroxidase–2/*Pox2–0, Pox2–1(Pox2–4C)*
Pox3	28	7	Peroxidase–3/*Pox3–1(Pox3–3C), Pox3–2(Pox3–5C)*
Pox4	28	7	Peroxidase–4/*Pox4–1, Pox4–2*
Pox5	28	6	Peroxidase–5/*Pox5–1, Pox5–2*

Contd...

Table 12.2–Contd...

Gene Symbol	Trait No.	Chromo-some No.	Name/Alelles/Gene Interaction/Mutagenesis[1]
*PoxcA**	30	3	Peroxidase–cA(cloned gene)
Pr(Rp)	1	4	Purple hull/somaclonal mutant
*Pro1**	12	12	Prolamin–1
*Pro2**	12	5	Prolamin–2
*Pro3**	12	5	Prolamin–3
*Pro4**	12	7	Prolamin–4
*Pro5**	12	5	Prolamin–5
Prpa(Pa)	1	1	Purple pericarp/complementary
Prpb(Pb)	1	4	Purple pericarp/somaclonal mutant
*PS**	19	6	Photoperiod sensitivity
Ps1	1	4	Purple stigma–1 (Gaisenmochi)/ expressed without *P*
Ps2(Ps1)	1	4	Purple stigma–2
*R5s1**	30	11	5S ribosomal DNA–1(5S rRNA–1)
*R5s2**	30	7	5S ribosomal DNA–2
*R5s3**	30	9	5S ribosomal DNA–3
R45s1(rDNA1)*	30	9	45S ribosomal DNA–1
R45s2(rDNA2)*	30	10	45S ribosomal DNA–2
*R45s3**	30	11	45S ribosomal DNA–3
*R45s4**	30	4	45S ribosomal DNA–4
Rc	3	7	Brown pericarp and seed coat/*Rc, Rc–s*
*RCH10**	30	3	Rice basic chitinase
rcn1	15	6	reduced culm number–1/gamma ray
rcn2	15	4	reduced culm number–2/EMS
rcn5	15	6	reduced culm number–5/gamma ray
Rd	3	1	Red pericarp and seed coat/ complementary with *Rc*
Rf1	24	10	Pollen fertility restoration–1/*Rf1, Rf1–a, Rf1–b, Rf1–c, Rf1–d, Rf1–e*
Rf2(Rfx)	24	2	Pollen fertility restoration–2
Rf3(R2,Rf2)*	24	1	Pollen fertility restoration for WA cytoplasm/duplicate

Table 12.2–Contd...

Gene Symbol	*Trait No.*	*Chromo-some No.*	*Name/Alelles/Gene Interaction/Mutagenesis*[1]
Rf4(R1,Rf1,RfWA1)*	24	7	Pollen fertility restoration for WA cytoplasm
Rf5(t)(Rf)*	24	1	Fertile revertant from II–32A, WA type CMS/gamma ray
*RfWA2**	24	10	Pollen fertility restoration for WA cytoplasm
rfs	4	7	rolled fine striped leaf/gamma ray, MNU
ri	14	5	verticillate rachis/spontaneous
rk1	11	4	round kernel–1/spontaneous
rk2	11	10	round kernel–2/ gamma ray
Rk3	11	5	Round kernel–3/spontaneous
rl1	16	1	rolled leaf–1/spontaneous, gamma ray
rl2	16	4	rolled leaf–2/spontaneous
rl3(rl1)	16	12	rolled leaf–3/spontanoous
rl4(rl2)	16	1	rolled leaf–4/gamma ray
rl5(rl3)	16	3	rolled leaf–5/MNU
rl6(t)	16	7	rolled leaf–6/unknown
S A1(A1)	20	6	Hybrid sterility–A/sporophytic F2 sterility/duplicate
S B2(B2)	20	6	Hybrid sterility–B
*Sc**	20	3	F1 pollen sterility–c/one locus sporogametophytic interaction
s a1(s1x1)	20	6	hybrid sterility–a1/gametophytic F1 sterility/duplicate
s c1	20	6	hybrid sterility–c1/gametophytic F1 sterility/duplicate
s c2	20	4	hybrid sterility–c2
s d1	20	6	hybrid sterility–d1/gametophytic F1 sterility/duplicate
s e1	20	3	hybrid sterility–e1/gametophytic F1 sterility/duplicate
s e2	20	4	hybrid sterility–e2
S1	20	6	Hybrid sterility–1/*S1, S1–a*/sporogametophytic interaction

Contd...

Table 12.2–Contd...

Gene Symbol	Trait No.	Chromo-some No.	Name/Alelles/Gene Interaction/Mutagenesis[1]
S3	20	11	Hybrid sterility–3/*S3*, *S3–a*/ sporogametophytic interaction
S5	20	6	Hybrid sterility–5/*S5–n*, *S5–j*, *S5–i*, *S5–p*/*S5–n*; wide compatibility
S6	20	6	Hybrid sterility–6/*S6*, *S6–a*/ sporogametophytic interaction
S7	20	7	Hybrid sterility–7/*S7–n*, *S7–kn*, *S7–ai*, *S7–cp*, *S7–j*
S8	20	6	Hybrid sterility–8/*S8–n*, *S8–i*, *S8–kn*, *S8–vp*
S9	20	7 or 4	Hybrid sterility–9/*S9–i*, *S9–kn*
S10	20	6	Hybrid sterility–10/*S10*, *S10–a*
S11(t)	20	11	Hybrid sterility–11/*S11*, *S11–a*
S13	20	1	Hybrid sterility–13/pollen killer
S15(S10)	20	12	Hybrid sterility–15/*S15–n* *S15–du*, *S15–i*
S16	20	1	Hybrid sterility–16/*S16–j*, *S16–n*, *S16–kn*
*S17(t)**	20	12	Hybrid sterility–17
*sal2**	30	6	salT protein
salT(sal1)*	29	1	salt tolerance
sbe1(QEI)*	12	6	starch branching enzyme–1 (Q enzyme 1)
*SBE III**	12	2	Starch branching enzyme III
sd1(d47)	8	1	dee-geo-woo-gen dwarf/*sd1*, *sd1–a*, *sd1–h*/spontaneous, gamma ray
sd7(t)	8	5	semidwarf–7(D56–31)/X ray
*sdg**	8	5	semidwarf (BRGPC)/spontaneous
Sdh1	28	12	Shikimate dehydrogenase–1/*Sdh1–1*, *Sdh1–2*, *Sdh1–3*, *Sdh1–4*
Se1(Lm,Lf,Rs,F1)	19	6	Photoperiod-sensitivity–1/*Se1*, *Se1–e*, *Se1–n*, *Se1–t*, *Se1–s*, *Se1–u*
se2	19	7	photoperiod-sensitivity–2
Se3(t)	19	6	Photoperiod-sensitivity–3
Se5	19	6	Photoperiod sensitivity–5/earliness due to the loss of photoperiod-sensitivity

Contd...

Table 12.2–Contd...

Gene Symbol	Trait No.	Chromo-some No.	Name/Alelles/Gene Interaction/Mutagenesis'
*Se8(t)**	19	7	Photoperiod-sensitivity–8
sepat(t)(se)*	19	6	photoperiod-sensitivity from Patpaku
sh1	11	11	shattering–1
sh2	11	1	shattering–2
Sh3	11	4	shattering–3
sh4(sb3)*	11	3	shattering–4
Shp1(Ex)	13	1	Sheathed panicle–1/spontaneous
Shp3(Ga)	13	5	Sheathed panicle–3/complementary with *Shp4*/spontaneous
Shp4(Gb)	13	3	Sheathed panicle–4/spontaneous
shp5(t)	13	4	sheathed panicle–5/MNU
shr1	12	1	shrunken endosperm–1/*shr1*, *shr1–s*, *shr1–a*/MNU
shr2	12	8	shrunken endosperm–2/MNU
sk2(t)(scl, fgr)	12	8	scented kernel–2
slg	11	7	slender glume/mutator/gamma ray
*sod**	30	3	superoxide dismutase
sp	13	11	short panicle
*Spk(t)**	15	9	Spreading stub
sp11(sl.bl2)	17	12	spotted leaf–1/MNU
sp12(bl3)	17	2	spotted leaf–2/MNU
spl3(bl4)	17	3	spotted leaf–3/gamma ray
spl4(bl5)	17	6	spotted leaf–4/gamma ray
spl5(bl6)	17	7	spotted leaf–5/gamma ray
spl6	17	1	spotted leaf–6/gamma ray
spl7	17	5	spotted leaf–7/gamma ray
spl8(bl8)	17	5	spotted leaf–8/MNU
spl9	17	7	spotted leaf–9/gamma ray
spl10	17	10	spotted leaf–10/MNU
spr1	13	4	spreading panicle–1
Spr3(t)	13	4	Spreading panicle–3

Contd...

Table 12.2–Contd...

Gene Symbol	Trait No.	Chromo-some No.	Name/Alelles/Gene Interaction/Mutagenesis'
Ssi1(Dm1)*	15	1	Short second internode–1 (dm-type)/X ray
*ssi2**	15	1	short second internode–2(dm-type)/ gamma ray
ssk(sk)	20	4	malformed semisterile/spontaneous
*Ssv**	15	7	Short stature (vegetative type)
st1(ws1)	4	6	stripe–1/spontaneous
st2(gw)	4	5	stripe–2/spontaneous
st3(st1)	4	3	stripe–3/MNU
st4(ws2)	4	4	stripe–4/gamma ray
st5	4	4	stripe–5/gamma ray
st6(t)	4	3	stripe–6/EMS
st7	4	4	stripe–7/spontaneous
st8	4	7	stripe–8/spontaneous
Stva(St1)	26	6	Stripe disease resistance–a/ complementary
Stvb(St2)	26	11	Stripe disease resistance–b/*Stvb*, *Stvb–i*
Sub1	15	9	Submergence tolerance–1
sug(su)	12	8	sugary endosperm/MNU
*Telsa1**	30	9	Telomeres
*Telsm1**	30	8	Telomeres–1
*Telsm3**	30	11	Telomeres–3
tms1(TGMS)	20	8	thermosensitive male sterility–1/ spontaneous
tms2	20	7	thermosensitive male sterility–2/gamma ray
tms3(t)	20	6	thermosensitive male sterility–3/gamma ray
*Tos1**	30	1	Retrotransposon–1
*tpi**	30	1	triosephosphate isomerase (cloned gene)
tri	11	2	triangular hull/spontaneous
TRYP(tin)*	30	1	Trypsin-inhibitor

Contd...

Table 12.2–Contd...

Gene Symbol	Trait No.	Chromo-some No.	Name/Alelles/Gene Interaction/Mutagenesis'
tsa	15	1	twisted stem–a/complementary/unknown
tsv1(RTSV)	26	4	rice tungro spherical virus resistance–1
Una	11	6	Uneven grain–a/complementary/spontaneous
Unb	11	7	Uneven grain–b/spontaneous
Ur1	14	6	Undulate rachis–1/spontaneous
ur2	14	8	undulate rachis–2/MNU
v1	4	3	virescent–1/spontaneous
v2	4	3	virescent–2/spontaneous
v3	4	6	virescent–3/MNU
v4	4	11	virescent–4/gamma ray
v5	4	3	virescent–5/MNU
v6	4	1	virescent–6/MNU
v7	4	3	virescent–7/MNU
v8	4	8	virescent–8/MNU
v9	4	11	virescent–9/MNU
v10	4	5	virescent–10/MNU
v11	4	7	virescent–11/MNU
Wc(=S5–n,S–n)*	20	6	Wide compatibility
*wgl(t)**	11	5	spikelet width/*wgl–n(t)*
Wh	3	4	White hull/spontaneous
wp1	3	7	white panicle–1/spontaneous
Wph1(Wbph)	27	7	Whitebacked planthopper resistance–1
Wx,(am)	12	6	glutinous endosperm/*wx, wx–op*; Waxy protein/*Wx–a, Wx–b*/spontaneous, MNU
Xa1(Xe)	25	4	Xanthomonas campestris pv. oryzae resistance–1 /*Xa1, Xa1–h*
Xa2	25	4	Xanthomonas campestris pv. oryzae resistance–2
Xa3(=Xa4b,Xa6, Xa9 Xaw)	25	11	Xanthomonas campestris pv. oryzae resistance–3
Xa4(Xa4a)	25	11	Xanthomonas campestris pv. oryzae resistance–4

Table 12.2–Contd...

Gene Symbol	Trait No.	Chromo-some No.	Name/Alelles/Gene Interaction/Mutagenesis[1]
xa5	25	5	Xanthomonas campestris pv. oryzae resistance–5
Xa7	25	6	Xanthomonas campestris pv. oryzae resistance–7
Xa10	25	11	Xanthomonas campestris pv. oryzae resistance–10
Xa12(Xakg)	25	4	Xanthornonas campestris pv. oryzae resistance–12/*Xa12, Xa12–h(Xa kg–h)*
xa13	25	8	Xanthomonas campestris pv. oryzae resistance–13
Xa14	25	4	Xanthomonas campestris pv. oryzae resistance–14
*Xa21**	25	11	Xanthomonas campestris pv. oryzae resistance–21
*Xaa**	25	11	Xanthomonas campestris pv. oryzae resistance–a
*Xah**	25	11	Xanthomonas campestris pv. oryzae resistance–h
*Xa(t)**	25	11	Xanthomonas campestris pv. orizae resistance
*xak**	25	7	Xanthomonas campestris pv. oryzae resistance–k
*ygl**	4	10	yellow green leaf/unknown
ylb	4	5	yellow banded leaf blade/unknown
ylm	4	4	yellow leaf margin/gamma ray
z1	4	11	zebra–1/spontanous
z2	4	11	zebra–2/gamma ray, MNU
z3	4	3	zebra–3/gamma ray, MNU
z4	4	8	zebra–4/MNU
z5	4	4	zebra–5/EI
z6	4	7	zebra–6/somaclonal mutant during anther culture
z7	4	5	zebra–7/spontaneous
z8	4	1	zebra–8/gamma ray

Contd...

Table 12.2–Contd...

Gene Symbol	Trait No.	Chromosome No.	Name/Alelles/Gene Interaction/Mutagenesis[1]
z9	4	3	zebra–9/unknown
z10	4	7	zebra–10/unknown
z11	4	2	zebra–11/unknown
z12	4	2	zebra–12/unknown
z13	4	6	zebra–13/unknown
*ZB8**	30	5	Phenylalanine ammonia lyase
zn	17	6	zebra necrosis/ gamma ray

(1) Name of chemical mutagen; EI: ethyleneimine, EMS: ethyl methanesulfonate, MNU: N-methyl-N-nitrosourea.

(2) ( ) means the old gene symbol or equivalent with the designated symbol.

* Gene symbols are not registered in the committee of Rice Genetics Cooperative.

(3) As to the literatures, refer to the tables mentioned in references [22, 23 and 24]. The database is also available at http://www.grs.nig.ac.jp in Rice Genetic Resources in Japan 1997 [24].

There still exist several problems about naming and usages such as genes for heading or flowering date (E, Ef and lh), sterility and lethality (S, s, D, L etc)., male sterility (classification and allelism tests), name of seed proteins (Glu, glup, Pro, etc)., isozymes (naming of cloned genes) and designation of molecular markers and QTLs (Kinoshita 1995).

Linkage Maps

According to Nagao [44], the first reliable linkage relation was established between C (Brown apiculus color) and wx (glutinous endosperm) with recombination values of 20•'22•" (Chao1928, Takahashi 1923, Yamaguchi 1926). Later, Nagao and Takahashi (Yamaguchi 1926) were the first to construct 12 linkage groups. Cytogenetical approaches using reciprocal translocations and primary trisomics have been developed to investigate relationships between linkage groups and individual chromosomes (Iwata and Omura 1971a, b, Sato *et al.*, 1973, Yoshimura 1982). Iwata and Omura (1984) and Khush *et al.* (1984) published independently revised linkage maps based on trisomic analyses.

Table 12.3: Coordinated Chromosome Numbering System and Its Comparison with Old Systems

New System Agreed	*Linkage Groups Kinoshita [20]*	*Trisomics* Khush et al. *[18]*	*Trisomics* *Iwata & Omura [14]*	*Chromosomes Kurata* et al.[a] *(1981)*	*Trans-Locations Nishimura*[b] *(1961)*
1	III	1	O	K1	3
2	X	2	N	K2	8
3	XI + XII	4	M	K3	5
4	II	12	E	K4	11
5	VI + IX	5	L	K9	2
6	I	3	B	K6	6
7	IV	7	F	K11	10
8	*sug*	8	D	K7	12
9	V, VII	9	H	K10	1
10	*fgl*	10	C	K12	7
11	VIII	11	G	K8	9
12	*d–33*	6	A	K5	4

[a] Quoted from Jpn. J. Genet. 56(1): 41–50, 1981.

[b] Quoted from Bull. Natl. Inst. Agr. Sci. Ser. D9: 171-235, 1961.

On the other hand, Misro *et al.* (Misro 1981) presented linkage maps based on the data of indica rice. However, it is difficult to identify different linkage groups because of a discrepancy in genic suppositions on anthocyanin coloration and related characters.

Rice chromosomes are difficult to distinguish from each other because of their small size and lack of characteristics, such as banding patterns and centromere positions. To avoid confusion among researchers, a rule was set up by RGC, indicating that chromosomes were assigned Arabic numerals in descending order of their pachytene length (or centromere position in case of ambiguity of length). As a result of joint observations and intensive discussions on the extra chromosome of primary trisomic plants, a unified system of numbering rice chromosomes was accepted during the Second International Rice Genetics Symposium in 1990. The agreed system of numbering rice chromosomes and linkage groups is shown in

Table 3. and the numbers of linkage groups trisomics, translocations and karyotypes follow this system (Khush and Kinoshita 1991, Kinoshita 1995).

The construction of current linkage maps is presented as shown in Fig. 12.1 and Table 12.4. In the linkage maps, the Kosambi mapping function was used to correct the recombination values between genes. In the table, recombination values are expressed as cM or percentages. Thus 552 genes were allotted to the twelve groups and 195 marker genes were positioned on twelve chromosomes.

Saturated molecular linkage maps are already constructed in several groups [Causse *et al.*, 1994,. Kurata *et al.*, 1994a, Maheswaran *et al.*, 1997, McCouch *et al.*, 1998, Nagamura *et al.*, 1947, Saito *et al.*, 1947) and the integration of the different linkage maps has been progressed (Kishimoto 1993, Xiao 1992, Yoshimura 1997). Comparative mapping and construction of synteny maps are presented among different cereals and grasses (Ahn 1993, Devos and Gale 1997, Kurata *et al.*, 1994b, Moore 1997, Van Deynze 1995). Recently the positions of centromeres and arm locations of RFLP markers on the molecular linkage maps were detected through gene dosage analysis using the secondary and telotrisomics (Khush *et al.*, 1996, Singh *et al.*, 1996a, b). Depending on the information the orientation of the maps and allocation of genes to short and long arms were resolved. Present linkage maps (Fig. 12.1) followed this orientation.

We need to intensify cooperative efforts to locate the unlocated genes to respective chromosomes and map those which have been assigned to the linkage groups. Some of the linkage groups (Fukui 1996, Fukui *et al.*, 1994,. Hiei *et al.*, 1997, Iwata and Omura 1971a) are still sparsely populated with marker genes. Efforts should be made to find additional markers for these linkage groups (Iwata and Omura 1971a, Kinoshita 1997, Kinoshita and Takahashi 1991).

New linkage maps that integrate conventional and biochemical markers such as isozyme, RFLP, RAPD and cloned genes are highly valuable for future work in genetics and breeding. For this purpose, linkage data related to isozyme genes which can be commonly located in conventional and molecular maps, may be used effectively for the construction of integrated maps (Kinoshita and Takahashi 1991). Continued efforts are needed to integrate all existing markers into one unified, densely populated map.

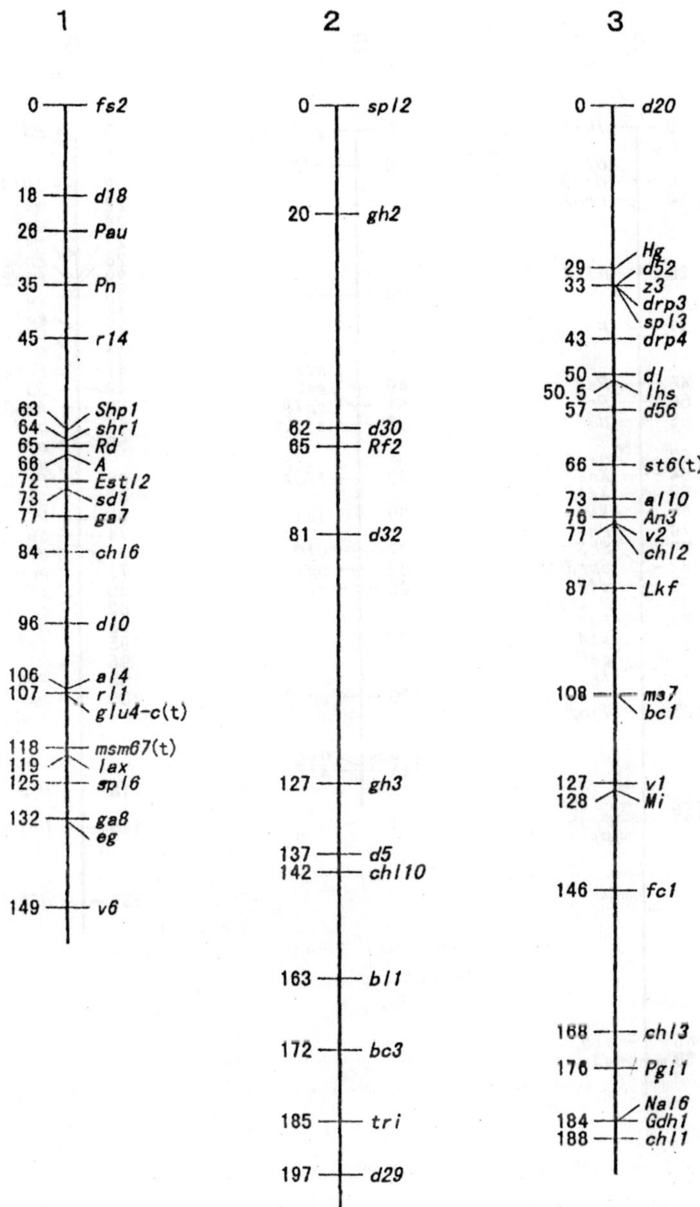

Fig. 12.1: Current Linkage Maps

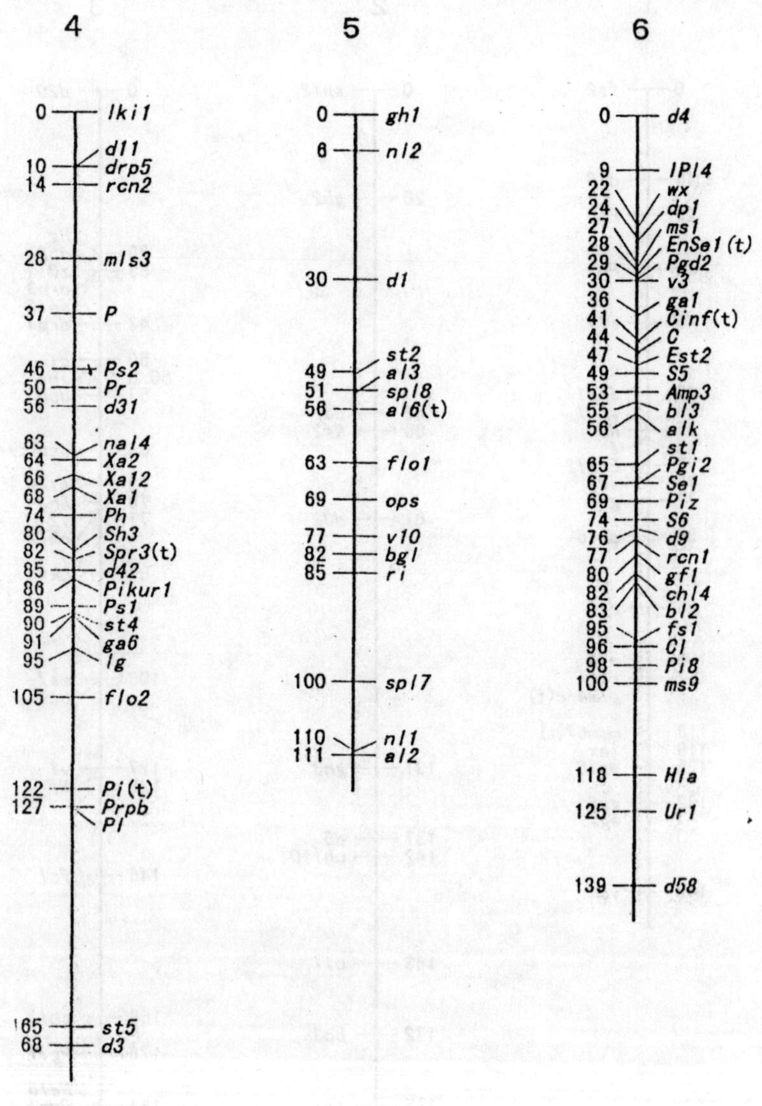

Fig. 12.1: Current Linkage Maps (Contd...)

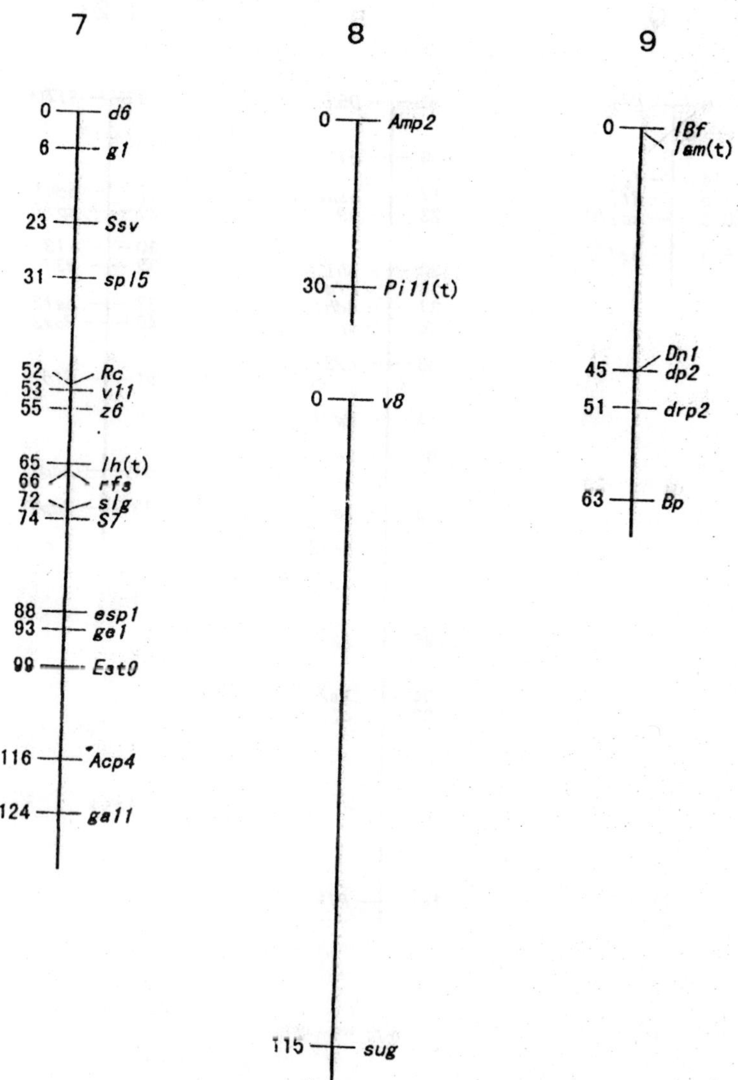

Fig. 12.1: Current Linkage Maps (Contd...)

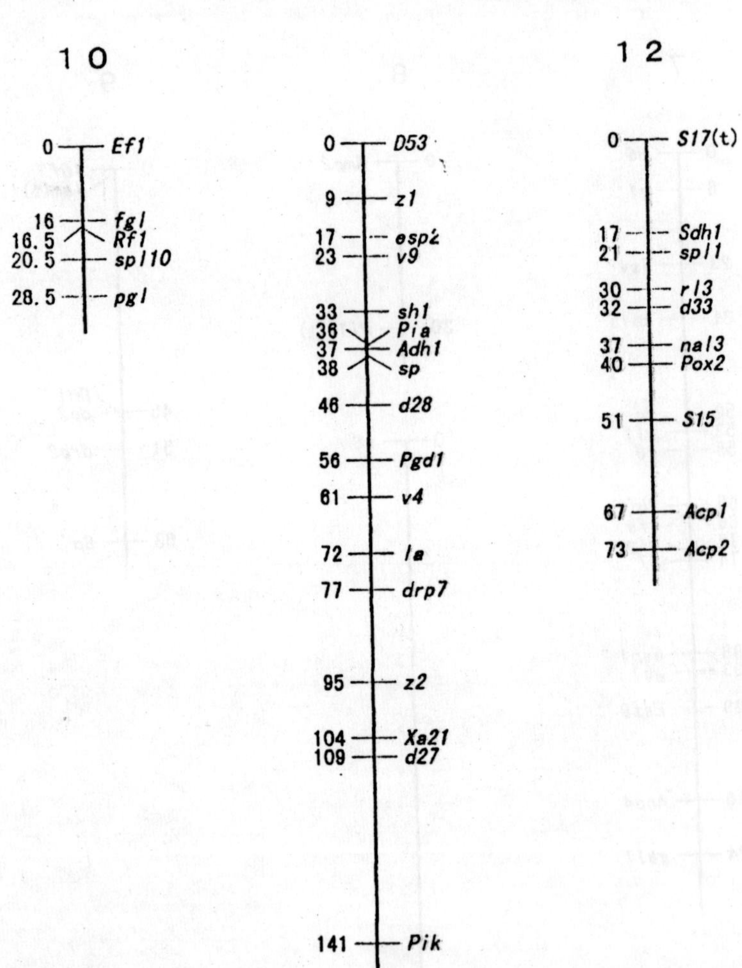

Fig. 12.1: Current Linkage Maps (Contd...)

Table 12.4: List of the Linkage Relations in the Respective Linkage Group

Chromosome	Gene	Locus and Linkage Relations
1	*fs2*	0, 8.7cM to XNpb216
1	*d18(d25)*	18, 4.1cM to XNpb96
1	*Pau*	26
1	*Pn*	35
1	*rl4(rl2)*	45, 9.4cM to XNpb368
1	*Shp1(Ex)*	63
1	*shr1*	64
1	*Rd*	65
1	*A(Sp)*	66, 4.4cM to XNpb121
1	*Eatl2**	72, OcM to RG220, near RZ92
1	*sd1(d47)*	73, near CDO962, 2.5cM to XNpb363, 0.8cM to RG220 and *Estl2,* 0.3cM to RG220 and to OPV10–500
1	*ga7*	77
1	*chl6*	84
1	*d10(d15,d16)*	96, near RG350 and RG462
1	*al4*	106
1	*rl1*	107
1	*glu4(t)**	107
1	*msm67(t)*(ms18)*	118
1	*lax(lx)*	119, 6.6cM to XNpb113, near CDO251
1	*spl6*	125, 5.4cM to XNpb174
1	*ga8*	132
1	*eg*	132, 5.3cM to XNpb97, bracketed by R2657A and L543 within 2.7cM
1	*v6*	149
1	*al8*	11 per cent–*d18–k*
1	*Ald2*(ald2)*	1.4cM to XNpb113, near C117
1	*Amy1B*(RAmy1B)*	1.0cM to RGI46X, near RZ274 and RZ276
1	*Arf2**	2.1cM–*Oc1**, 3.0cM to XNpb370
1	*chl5*	13.2 per cent–*eg*

Contd...

Table 12.4–Contd...

Chromosome	Gene	Locus and Linkage Relations
1	*d2*	15.6cM to XNpb359
1	*d26*	37 per cent–*A*
1	*d54(dK5)*	25 per cent–*lax*
1	*d55(dK6)*	15 per cent–*eg*
1	*Est5*	triplo–1
1	*Est10*	12.6cM to RG220
1	*ga9*	13 per cent–*d18–k*
1	*gf2*	triplo–1, 21cM–z8
1	*Glu1(Srpl,Gt2)*	17 per cent–*sd1*
1	*Glu1t* (Glu1)*	3.0 per cent to R758A
1	*Glu2t*(Glu2)*	3.0 per cent to R758A
1	*Glu3t*(Glu3)*	3.0 per cent to R758A
1	*glu3(t)**	10.2cM to C250
1	*Got1*	triplo–1
1	*led1*	triplo–1
1	*IPsb(IPs2)*	linked with *A*
1	*lgt*	16 per cent–*d26*
1	*mp1*	triplo–1
1	*msm77(t)**	15 per cent–*mp1*
1	*Oc1**	1.lcM to XNpb92
1	*Pit*	near G359, 0cM to R1613, 2.0cM to C470
1	*Prpa(Pp)*	7.2 per cent–*A*
1	*Rf3*(Rf2)*	1.4cM~2.6cM to RG532, 0cM~5.3cM to OPK05–800, 1.9cM to RG532, linked to K5 and W1
1	*Rf5(t)*[Rf(t)]*	5.3cM to OPB07–640 and OPBI8–1000, bracketed by RG374 and RG394
1	*S13*	near *A*
1	*S16*	20 per cent~29 per cent–*Est5*
1	*salT*(sal1)*	0.9cM to CDO548
1	*sh2*	11 per cent–*sd1*, near XNpb174

Contd...

Table 12.4–Contd...

Chromo-some	Gene	Locus and Linkage Relations
1	*Ssi1*(Dm1)*	39 per cent–*d18–k*, 10.1 per cent–*shr1–s*, 21.2 per cent–T(1)
1	*ssi2**	26.9 per cent–*Ssi1*, 1.8 percent–T(1)
1	*Tos1**	26%–*sd1*
1	*tpi**	near C146
1	*TRYP*(tin)*	near RG472 and RZ543
1	*tsa*	23 per cent–*A*
1	*z8*	triplo–1
2	*spl2(bl3)*	0, 9.2cM to XNpb349
2	*gh2*	20
2	*d30(dw)*	62, near XNpb227, near RG256
2	*Rf2(Rfx)*	65
2	*d32(dk4,d12)*	81
2	*gh3*	127
2	*d5*	137, near RG256
2	*chl10*	142
2	*bl1*	163, 10.3cM to XNpb250
2	*be3*	172
2	*tri*	185, 16.7cM to XNpb250
2	*d29(dK1)*	197
2	*ae3(t)**	triplo–2
2	*Amp1(Lap1,LapF)*	triplo–2, 19.4cM to XNpb223
2	*Amy1A/C*(RAmy1A/C)*	9.2cM to RG95, near RG151 and RG73
2	*ant**	——
2	*be5*	triplo–2
2	*eye2**	near RZ957
2	*D1**	39 per cent–*Ampl*
2	*Efx(t)**	18 per cent–*TR2–9*
2	*Glu4t*(Glu4)*	0.7 per cent to *Glu5t**
2	*Glu5t*(Glu5)*	1.7 per cent to XNpb227
2	*Glu6t*(Glu6,GluB,C,F)*	0 per cent to XNpb21

Table 12.4–Contd...

Chromo-some	*Gene*	*Locus and Linkage Relations*
2	*Glu7t*(Glu7,GluA)*	0 per cent to XNpb21
2	*glu1(t)*(Lge1)*	near XNpb243
2	*Got3*	triplo–2
2	*gpd1**	near C1061
2	*ms17*	35 per cent–*gh2*
2	*Pal1**	close to RZ318
2	*Pib(Pis)*	5.8 per cent–TR2–10, 0.5cM to GN234, 0cM to R1792, bracketed by C2782B and C379 within 3.6cM, bracketed by RZ213 and G1234, 0cM to S1916
2	*Pi14(t)*	34.3 per cent–*Amp1*
2	*Pi16(t)*	37.9 per cent–*Amp1*
2	*SBEIII**	1.0cM to RG157
2	*Z11*	triplo–2
2	*z12*	triplo– 2
3	*d20*	0
3	*Hg*	29, near RG348 and CDO20
3	*d52(dK2)*	33
3	*z3*	33
3	*drp3*	33
3	*spl3(bl4)*	33, 15.0cM to XNpb173
3	*drp4*	43
3	*dl(lop)*	50, 16.6cM to XNpb394
3	*lhs(lhs2,op)*	50.5
3	*d56(dK7)*	57
3	*st6(t)*	66
3	*al10*	73
3	*An3*	76
3	*v2*	77
3	*chl2*	77
3	*Lkf*	87
3	*ms7*	108

Contd...

Table 12.4–Contd...

Chromo-some	Gene	Locus and Linkage Relations
3	*be1*	108
3	*v1*	127
3	*Mi*	128
3	*fe1*	146, 10.9cM to XNpb249
3	*chl3*	168
3	*Pgi1(Pgia)*	176, near C217, bracketed by G232 and C217, near CDO87
3	*Na16*	184
3	*Gdh1*	184
3	*chl1(ch1)*	188, 7.1cM to XNpb164
3	*ACC1**	near RG510
3	*Actin1*(act1)*	near CDO337
3	*bl4*	29 per cent–*bc1*
3	*CatA1**	1.4cM to XNpb279
3	*cdc**	—
3	*d14(d10)*	32 per cent–*dl*
3	*E3*	37.5 per cent–*bc1*
3	*ga2*	11 per cent–*d1*
3	*ga3*	34 per cent–*d1*
3	*ga14*	bracketed by *Pgi1* and *bc1* within 0.2 per cent
3	*Grh(t)*(Grha)*	0 per cent to XNpb144
3	*Hd6(t)**	bracketed by R2311 and R1618 within 1.0cM
3	*L**	0.29cM to XNpb23, 0.8cM to XNpb23
3	*PHYA**	0.9cM to RM203
3	*pms2**	near RG191, linked to pRD10A
3	*PoxcA**	1.1cM to XNpb244–2
3	*RCH10**	near RG913X
3	*rl5(rl3)*	13 per cent–*chl1*
3	*Sc**	0.5cM to RG227
3	*s e1*	16 per cent–*bc1*
3	*sh4*(sh3)*	near XNpb40

Contd.

Table 12.4–Contd...

Chromo-some	*Gene*	*Locus and Linkage Relations*
3	*Shp4(Gb)*	27 per cent–*bc1*
3	*sod**	—
3	*st3(st1)*	1.1 per cent–*bc1*
3	*v5*	2.0 per cent–*ch1*
3	*v7*	1.7 per cent–*bc1*
3	*z9*	13cM–*Pgi1*
4	*lki1*	0
4	*dl1(d8)*	10, 11.0cM to XNpb49 near CDO456
4	*drp5*	10
4	*rcn2*	14
4	*mls3*	28
4	*P(Pa,Pb,Pc)*	37
4	*Ps2(Ps1)*	46
4	*Pr(Rp)*	50, 3.5cM–*Prpb,* 16.8cM to RG163, near RG177X and RZ939
4	*d31*	56
4	*nal4(nal)*	63
4	*Xa2*	64, 2.0cM to XNpb264, near RG476
4	*Xa12(Xakg)*	66
4	*Xa1(Xe)*	68, OcM to Y03–700, 1.5cM to U08–750, near CDO539X, 0cM to XNpb235 and C600
4	*Ph(=Bhe, Po)*	74, 0.1cM to V17, 1.5cM to XNpb261, 2.2cM to XNpb267, 5.9cM to RG214 bracketed by V17 and Ky3 within 0.1cM
4	*Sh3*	80
4	*Spr3(t)*	82
4	*d42*	85
4	*Pikur1**	86
4	*Ps1*	89
4	*st4(ws2)*	90
4	*ga6*	91
4	*lg*	95, 3.4cM to XNpb197, 9.2cM XNpb120

Contd

Table 12.4–Contd...

Chromo-some	Gene	Locus and Linkage Relations
4	*flo2*	105
4	*Pi(t)**	122, near CDO539x
4	*Prpb(Pb)*	127, 19.0cM to RG329
4	*Pl*	127
4	*st5*	165
4	*d3*	168
4	*al5*	34 per cent–*lg*
4	*al7(t)*	31 per cent–*lg*
4	*An1*	5.4 per cent–*d11*
4	*aul*	triplo–4
4	*Bph3*	—
4	*bph4*	close to *Bph3*, near C891
4	*Cht4*(Chi4)*	0cM to R1849, 0.3cM to C1399
4	*drp1*	39 per cent–*d2*(chr1)
4	*drp8*	28 per cent–*lg*
4	*ga10(t)*	27 per cent–*lg*
4	*ga12*	2.1 per cent–*shp5(t)*
4	*Glh3*	34 per cent–*bph4*
4	*Glh(t)**	2.1cM to RZ262 and OP246
4	*glup3(esp7)*	trisomic–4, triplo–4
4	*Gm2*	1.3cM to RG329, near RZ590, near $F8_{1700}$ and $F10_{600}$
4	*gpd2**	near C377
4	*hpg1**	13.3cM to L102
4	*nal1*	25 per cent–*d2*(chr. 1)
4	*nal5(nal1)*	9.5 per cent–*lg*
4	*ocp**	—
4	*Pi5(t)*	bracketed by RG498 and RG788, near RZ565, or 5.9 per cent to G103(chr. 9)
4	*Pin1*	31 per cent–*P1*
4	*R45s4**	*in situ* hybridization
4	*rk1*	35 per cent–*lg*

Contd...

Table 12.4–Contd...

Chromosome	*Gene*	*Locus and Linkage Relations*
4	*rl2*	35 per cent–*d2* (chr. 1)
4	*s c2*	31 per cent–*Ph*
4	*s e2*	15 per cent–*lg*
4	*shp5(t)*	near *d2*(chr. 1)
4	*spr1*	27 per cent–*Pl*
4	*ssk(sk)*	6.8 per cent–*Pl*
4	*st7*	triplo–4
4	*tsv1(RTSV)*	4.5cM to Rz262 and OP246
4	*Wh*	8.0 per cent–*lg*
4	*Xa14*	—
4	*ylm*	10 per cent–*lg*
4	*z5*	11 per cent–*lg*
5	*gh1*	0
5	*nl2*	6
5	*d1*	30 near XNpb105, 0.8cM to C309 and S2351
5	*st2(gw)*	49
5	*al3*	51
5	*spl8(bl8)*	51
5	*al6(t)*	56
5	*flo1*	63
5	*ops*	69
5	*y10*	77
5	*bgl*	82
5	*ri*	85
5	*spl7*	100, near XNpb81
5	*nl1*	110, 0cM to XNpb188
5	*al2*	111
5	*ACC3**	near RZ495
5	*Ald1**	0.8cM–*Cht2*, 7.7cM to C13
5	*An2*	33 per cent–*gl1*
5	*Arf1**	1.7cM to *Oc2*

Contd...

Table 12.4–Contd...

Chromo-some	Gene	Locus and Linkage Relations
5	*ATP*(atp1)*	4.7cM to RG435, near RG697
5	*bc2*	triplo–5, 29cM–*spl7*
5	*bd1*	22 per cent–*gl1*
5	*CALa**	near RG480
5	*Cht2**	0.8cM–Ald1*, 2.9cM to XNpb105
5	*Em**	trisomic–5
5	*er(o)*	38 per cent–*gh1*
5	*Est13*	17.9cM to RG13
5	*eui*	27 per cent–*nl1*
5	*gl1*	12 per cent–TR1–5d, 10.5cM to XNpb396, 14.3cM to RG182
5	*Glh6*	triplo–5
5	*glup4(esp8)*	trisomic–5
5	*Grh1*	OcM to C309 and XNpb260
5	*IPl1*	31 per cent–*gh1*
5	*Mdh3**	MAAL5
5	*MPox1 (MPx1)*	near *gl1*
5	*ms14*	11 per cent–*nl1*
5	*Oc2**	0.4cM to XNpb255
5	*Pi10(t)**	2.9cM to RRHI8, bracketed by OPFC2700 and OPH18–2400 within 7cM
5	*Pox1(Px,Pe)*	38 per cent–nl1
5	*Pro2**	2.2 per cent–*Pro3**, 6.8 per cent to XNpb251
5	*Pro3**	2.2 per cent–*Pro2**, 7.8 per cent to XNpb251
5	*Pro5**	3.0 per cent–*Pro3**, 7.8 per cent to XNpb251
5	*Rk3*	triplo–5
5	*sd7(t)*	0 per cent–d1
5	*sdg**	4.3cM to RZ182
5	*ShP3(Ga)*	37 per cent–*nl1*
5	*wgl(t)*	OcM to W168A, Y1060L and Y1060R
5	*xa5*	trisomic–5, near, *gl1*, 0.8cM to RG207, near XNpb396, near RG556A, OcM to RZ390, RG556 and RG207

Contd...

Table 12.4–Contd...

Chromo-some	*Gene*	*Locus and Linkage Relations*
5	*ylb*	32 per cent–*nl1*
5	*z7*	triplo–5
5	*zB8**	near RZ182
6	*d4*	0
6	*lpl4*	9
6	*wx(am)*	22, 1.lcM to C425A, near RZ1002, near XNpb209
6	*dp1*	24, 10.8cM to XNpb209
6	*ms1(sf)*	27
6	*EnSe1(t)*	28
6	*Pgd2*	29
6	*v3*	30
6	*ga1*	36
6	*Cinf(t)*(Ler)*	41
6	*C*	44, 16.4cM to XNpb165–1
6	*Est2*	47, 1.7 per cent to RG213, 14cM to RZ450, near RZ144 and RZ667
6	*S5*	49, 1.6cM-;*Est2*, 4.1cM-Amp3 and Est2, 4.4cM to RG213
6	*Amp3(Amp1)*	53, near RZ144, 1.7 per cent to RG213
6	*bl3*	55
6	*alk*	56, 4.6cM–*Pgi2,* 1.1cM to C1478
6	*st1(ws1)*	65
6	*Pgi2(Pgih)*	67, 4.6cM-alk, 7.9 per cent to RG213, near hmg2, 7.6cM to, RZ144, near RRA–19, near RZ667
6	*Se1(Lm,Lf,RS,Fl,Hd1)*	67, 1.3cM–*Piz–t,* 4.7 per cent to XNpb294, near RZ612 and BRA–19, 4.9cM to XNpb233, 0.3cM to C235, near S2539 and R1679
6	*Piz(Pi2)*	69, 2.9cM to XNpb294, 2.1cM to RG64, 2.8cM to RG64, 5.0cM to RG456
6	*S6*	74
6	*d9*	76
6	*ren1*	77

Contd...

Table 12.4–Contd...

Chromosome	Gene	Locus and Linkage Relations
6	*gf1*	80
6	*chl4*	82
6	*bl2(blm)*	83
6	*fs1*	95
6	*Cl*	96, near RG172
6	*Pi8*	98
6	*ms9*	100
6	*Hla*	118
6	*Ur1*	125
6	*d58*	139
6	*al1(alK1)*	7.1 per cent–*wx*
6	*al9(t)*	trisomic–6
6	*Amp5*	15.0cM–*Est2*
6	*Amy2A*(RAmy2A)*	3.7cM to RZ884, 6.3cM to RG433
6	*aph*	linked with *Est2* and *Pgi2*
6	*bc4*	triplo–6
6	*Cat1*	22 per cent–*Pox5*, 8.1cM to RG244, linked to RG433
6	*chl7*	27 per cent–*Piz*, 13 per cent–*Pgi2*
6	*Cht1(Chi1)*	0.4cM to XNpb342, near CDO218
6	*Cht3**	0.4cM to XNpb342
6	*d21*	8.3 per cent–*wx*
6	*drp6*	15 per cent–*fc2*
6	*du2035**	triplo–6
6	*duEM47**	triplo–6
6	*dw1 (fh1)*	30 per cent–*Se1*
6	*Efy(t)**	9.2 per cent–T6–7
6	*Enp1**	2.2 per cent–*Cat1*
6	*fc2*	18 per cent–*C*
6	*ga4(gaA)*	34 per cent–*wx*
6	*ga5(gaB)*	27 per cent–*wx*

Contd...

Table 12.4–Contd...

Chromosome	Gene	Locus and Linkage Relations
6	*Glup6(t)**	trisomic–6
6	*Got2*	31 per cent–*Pgi2*
6	*Grh3(t)**	6.2 per cent to G2028, 0cM to C81
6	*Hd3(t)**	bracketed by R2967 and C1032 within 1.1cM, near C226A
6	*IPl2*	10 per cent–*Ipl4*
6	*Ldh**	0.9cM to C1003B
6	*mp2*	triplo–6
6	*Pi9(t)**	near RRA–19
6	*Pi13(t)**	36.9 per cent–*Amp3*
6	*Pi13**	0cM to C58 and R1960
6	*pms**	linked to pRD10B
6	*Pox5*	39–*Pgi2*
6	*PS**	23 per cent–*Pgi2*, 0cM to RG64
6	*ren5*	22 per cent–*C*
6	*S A1(A1)*	9.59 per cent–*C*
6	*S B2(B2)*	28 per cent–*wx*
6	*s a1(s1,x1)*	21 per cent–*wx*
6	*s c1*	8.6 per cent–*C*
6	*s d1*	33 per cent–*wx*
6	*S1*	near *C*
6	*S8*	11 per cent–*Cat1*
6	*S10*	0.15cM–*wx*
6	*sal2**	—
6	*sbe1*(QEi)*	D10752, 0.3cM to G342
6	*Se3(t)*	near CDO17 and CDO96
6	*Se5*	27 per cent–*Piz–t*, 23cM–*Pgi2*
6	*sepat(t)*(se)*	14 per cent–*wx*, 13 per cent–*C*
6	*spl4(bl5)*	2.5 per cent–*dp1*, 5.0cM to XNpb209
6	*Stva(St1)*	38 per cent–*wx*
6	*tms3(t)**	7.7cM to OPAC3–640

Contd...

Table 12.4–Contd...

Chromo-some	Gene	Locus and Linkage Relations
6	*Una*	22 per cent–*Cla*
6	*Wc*(=S5–n,S–n)*	5.9cM–*Est2*, and 12.7cM to RG398
6	*Xa7*	8.8 per cent to G1091
6	*z13*	triplo–6
6	*zn*	20 per cent–*C*
7	*d6(d34)*	0, near XNpb50
7	*gl(1ng)*	6, 24.9cM to XNpb85–1
7	*Ssv**	23
7	*spl5(bl6)*	31, near XNpb85–1
7	*Rc*	52, 0cM to C67 A and R277, 12.5cM to RG30
7	*v11*	53, near XNpb85–1
7	*z6*	55
7	*lh(t)**	65
7	*rfs*	66, 1.4cM to XNpb20
7	*slg*	72, 3.0 per cent to XNpb33, 0.6cM to R1440
7	*S7*	74
7	*esp1(rsp1)*	88, 7.8cM to RZ394
7	*ge1*	93, 2.4cM to RZ395
7	*Est9(Earel)*	99, near RZ337B
7	*Acp4*	116, near RZ387 and RG417
7	*ga11*	124
7	*betaAmy¹*	trsomic–7
7	*CALb**	near RG351
7	*d7*	39 per cent–*d6*
7	*d60[sd(t)]*	trisomic–7
7	*ds3(t)*	triplo–7, acro–7
7	*El(=mEf1+)*	trisomic–7, 23 per cent–*Rc*
7	*clf2**	—
7	*Est1*	triplo–7
7	*Est7*	24 per cent–*Acp4*
7	*Hd2**	near C728

Contd...

Table 12.4–Contd...

Chromo-some	*Gene*	*Locus and Linkage Relations*
7	*Hd4**	near L538T3
7	*hwd2*	bracketed by G338 and C492 within 1.1cM
7	*lp1*	12 per cent–*Unb*
7	*Mal1*	29.3cM to RG528, 22.7 per cent to RG173 or 10.8 per cent–*lg*(chr. 4)
7	*Mal2*	1.8cM to RG511
7	*ms8*	20 per cent–*rfs*
7	*Pil7(t)*	linked to *Est9*
7	*pms1**	4.3cM to RG477, linked to RGMS0.7
7	*Pox3*	5 per cent–*Pox4*
7	*Pox4*	31 per cent–*Est1*
7	*Pro4**	1.6 per cent to XNpb338
7	*R5s2**	short arm
7	*Rf4* (Rf1, RfWA1)*	triplo–7, near XNpb379
7	*rl6(t)*	12 per cent–*lp1*
7	*S9*	16.2 per cent or 18.6 per cent–*Est1*
7	*se2*	23 per cent–*gl*
7	*Se8(t)**	0cM to S2267 and S1979
7	*spl9*	triplo–7
7	*st8*	4cM–*wp1*
7	*tms2*	bracketed by R643A and Rl440 within 1.7cM
7	*Unb*	18 per cent–*g1*
7	*wp1*	triplo–7
7	*Wph1(Wbph)*	near RG146A
7	*xak**	triplo–7
7	*z10*	0cM–*wp1*, triplo–7
8	*Amp2(AlapA)*	0, 8.3cM to RG136, near RG418B
8	*Pi11(t)*)Pizh)*	30, near RZ323 and RZ617, 14.9cM to Bp127
8	*v8*	0, near XNpb278
8	*sug(su)*	115, near XNpb397
8	*Amp4*	triplo–8
8	*Amy3D/E*(RAmy3D/E)*	0.9cM to RZ952, 11.1cM to RG1

Table 12.4–Contd...

Chromo-some	Gene	Locus and Linkage Relations
8	*An4*	5.0 per cent–TR7–8b, 0.85 per cent–TR8–12
8	*chl8*	triplo–8
8	*chl9*	triplo–8
8	*d51(dK8)*	trisomic–8
8	*gax**	linked to *An4*
8	*Hd5**	near R902
8	*pkc2**	—
8	*shr2*	trosomic–8
8	*sk2(t)(sel,fgr)*	2.1cM to RG28, 6.7 per cent to XNpb126 linked to *AG8*
8	*Telsm1**	near RG29X
8	*tsm1(TGMS)*	6.7cM to TGMSI.2(OPB–19)
8	*ur2*	trigomic–8
8	*xa13*	3.8cM to RG136, 3.7cM to RG136, 5.3cM to AC5–900
8	*z4*	trisomic–8
9	*IBf*	0, 11.1cM to XNpb36
9	*lam(t)*	0
9	*Dn1*	45, near RG553
9	*dp2*	45, 3.3cM to XNpb385
9	*drp2*	51, 2.9cM–*dp2*, 0cM to XNpb385, 10.6cM to XNpb339
9	*Bp*	63, 1.4cM to XNpb339
9	*Adh2**	trisomic–9
9	*Amy3A/B/C** (*RAmy3A/B/C*)	9.4cM to RZ228
9	*atub**	2.2cM to RG757
9	*Bph(t)**	31.9cM to RZ404
9	*chs1(t)**	27 per cent–*dp2*
9	*cyc1**	1.6cM to RZ792, 4.6cM to RZ404
9	*d57[d(x)]*	21 per cent–*Dn1*
0	*dp(t)**	10 por oont *IBf*

Contd..

Table 12.4–Contd...

Chromosome	Gene	Locus and Linkage Relations
9	*du2120**	triplo–9
9	*Est3*	16.4 per cent to XNpb257
9	*Est12*	29.1cM to RZ12
9	*Glup1(Esp5,Gup1)*	trisomic–9
9	*glup2(esp6,gup2)*	trisomic–9
9	*Glup5(t)**	trisomic–9
9	*gm1*	35 per cent–*IBf*
9	*hsp2**	—
9	*hsp4**	—
9	*Man*	4.3cM to XNpb293
9	*ms10*	5 per cent–*Dn1*
9	*pkc1**	—
9	*R5s3**	short arm
9	*R45s1*(rDNA1)*	4.6cM–*atub**, 1.0cM to XNpb315, 2.5cM to RG757X
9	*Spk(t)**	0cM to C570, 1.0cM to C506
9	*Sub1*	4cM to C1232, bracketed by C1232 and RZ698
9	*Telsa1**	8.8cM to R45s1*
10	*Ef1(Ef2)*	0
10	*fg1(f1)*	16, 0cM to XNpb291
10	*Rf1*	16.5, 7.5cM to XNpb291, near RG561 and RG134, 3.7cM to OSRRf5, 1.5 per cent to fL601, ncar fL601B
10	*spl10*	20.5, 10.9cM to XNpb127
10	*pgl*	28.5, 7.5cM–*spl10*, 3.5cM to XNpb127, near RG257 and RG134
10	*du1*	trisomic–10
10	*Ef(t)**	9.96cM to CDO98
10	*gew2**	—
10	*gew4**	—
10	*glu2(t)**	6.8cM to G1082
10	*hwd1*	close to L169, 0.6cM to R2309

Contd...

Table 12.4–Contd...

Chromo-some	Gene	Locus and Linkage Relations
10	*R45s2*(rDNA2)*	0cM to XNpb32
10	*RfWA2**	triplo-10
10	*rk2*	2.5 per cent–TR10–11, 7.5cM–pgl
10	*ygl**	7.5cM–*pgl*, triplo–10
11	*D53(DK3)*	0
11	*z1*	9
11	*esp2(rsp2)*	17
11	*v8*	23, 11.1cM to XNpb183
11	*sh1*	33
11	*Pia*	36
11	*Adh1(Adh1/2)*	37, 7.3 per cent to RG118, 13cM to RG1094, near RM120
11	*sp*	38, 7.7cM to XNpb320
11	*d28(dC)*	46
11	*Pgd1*	56, near RZ797 and CDO365
11	*v4*	61
11	*la*	72, 3.0cM to RG1094, 13.1cM to XNpb202
11	*drp7*	77
11	*z2*	95, 0cM to CDO365, 4.2cM to gmz410
11	*Xa21**	104, 2.2 or 2.7cM to pTA818, 0cM to RAPD818 and RG103, 0cM to RM21, 2.8cM to AB9
11	*d27(dt)*	109, 2.6cM to pTA818, near RZ141 and RZ537
11	*Pik*	141, 17.8 per cent to G181, 1.6cM to R1506, 0.5cM to R543, 1.5cM to L190
11	*adh2(t)**	—
11	*chs**	3.3cM to RC2
11	*D2**	37.3 per cent–*Pgd1*
11	*Dse**	linked to *Adh1* and *Pgd1*
11	*esp3(rsp3)*	trisomic–11
11	*Fdp1**	triplo–11
11	*gal(lt)*	24 per cent–*la*
11	*Grh2(Grhb)*	0 per cent to G1465, near G4001
11	*Lk2(t)(Lknb)*	19 per cent–*sp*

Contd...

Table 12.4–Contd...

Chromosome	*Gene*	*Locus and Linkage Relations*
11	*mik*	24 per cent–*sh1*, 38 per cent–*la*
11	*MPiz(Rb6)*	11 per cent–*la*
11	*nal2*	36 per cent–*la*
11	*OsFAD3**	10.4cM to RM20B
11	*OsFADY**	2.3cM to T28
11	*PBR**	10.8cM to CDO226
11	*Pbst**	13.8cM–*Stvb–l*, 9.4cM to ST723 and CDO226
11	*Pif*	15 per cent–*Pik*
11	*Piis1(Rb4)*	23 per cent–*la*
11	*Pikur2**	14 per cent–*la*
11	*Pise1(Rb1)*	9.5 per cent–*la*
11	*Pi1*	4.7 per cent–*Pik*, 3.5cM to NpB181, near RZ353X and RZ424, 14.0cM to RZ536
11	*Pi7(t)*	near CDO365, bracketed by RG103A and RG16
11	*Pi18(t)**	5.4cM to RZ536
11	*R5s1**	2cM to RG247X, near RG16 and RG167X
11	*R45s3**	*in-situ* hybridization
11	*S3*	1 per cent–*la*
11	*S11(t)*	6.2 per cent–*la*
11	*Stvb(St2)*	3.0cM to XNpb257 and XNpb254, near ST10
11	*Telsm3**	near RG304A
11	*Xa3(=Xa4b,Xa6, Xa9,Xaw)*	22 per cent–*d27*, 2.3cM to XNpb181, near RZ424
11	*Xa4(Xa4a)*	near–*Xa3*, 1.7cM to XNpb78, 1.7cM to XNpb181, 17cM to RG53, near RZ536
11	*Xa10*	27 per cent–*Xa4*, 5.3cM to $Oo7_{2000}$ near CDO520
11	*Xaa**	28 per cent–*la*, triplo–11
11	*Xah**	17 per cent–*Xaa**
11	*Xa(t)**	13 per cent–*Xa4*, 4.8cM to R543
12	*S17(t)**	0
12	*Sdh1*	17, 4.9cM to XNpb402, near RZ76, 6.7 per cent to XNpb860–C

Contd...

Table 12.4–Contd...

Chromo-some	Gene	Locus and Linkage Relations
12	*spl1(sl,bl2)*	21, 3.6cM to XNpb88, 4.3cM to XNpb154
12	*rl3(rl1)*	30, near XNpb402
12	*d33(dB)*	32, 7.5cM to XNpb148
12	*nal3(nal2)*	37
12	*Pox2*	40, near CDO459
12	*S15*	51
12	*AcP1*	67, 0cM–*Acp2*, near RG176X, 29.2cM to RG361
12	*AcP2*	73, 1.9cM to RG181A
12	*Ald3*(ald1)*	near C1336
12	*Bph1*	10.9cM to XNpb248, near C148 and C87, 5.0cM to C185, 10.9cM to XNpb248
12	*bph2*	30cM–*BPh1* near G2140
12	*Bph9*	near C449
12	*Bph10(t)*	3.68cM to RG457
12	*du4*	trisomic–12, 31 per cent–*d33*
12	*ga13*	10 per cent–*Acp1*
12	*Hbv*	near RG121X
12	*IPi(t)**	near RZ670
12	*IPi3(t)**	near RZ670
12	*Lec**	near RG190 and RG323
12	*OsFADX**	5.7cM to RM20A
12	*Pita(sl,Pi4a)*	5 per cent–*spl1*, 0cM to XNpb289 and XNpb239, 3.3cM to RZ397, 15.4cM to RG869
12	*Pi6(t)**	near RG81
12	*Pi12**	5.1cM to RG869
12	*Pil4**	close or allelic to *Pita–2*
12	*Pi62(t)**	0cM to SP7C3 and SP8C6
12	*Pi157**	3.0cM to RG341
12	Pro1*	3.2 per cent to XNpb88

( ) means the old gene symbol or equivalent with the designated symbol.

* Gene symbols are not registered in the committee of Rice Genetics Cooperative.

As to the molecular markers, refer to the literatures [2, 27, 51 etc.]

The utility of biochemical and molecular markers is based on finding tight linkages between RFLP markers and the target gene. Major genes tagged with molecular markers hitherto in rice linkage groups are listed in Table 12.4.

Once a gene is tagged with RFLP markers, it can easily be transferred to other varieties by indirect selection with the RFLP markers in the progeny from an appropriate cross. Tagging genes for blast resistance, bacterial leaf blight resistance and insect resistance may ultimately lead to marker-aided selection and the cloning of those genes via chromosome walking (marker-based gene cloning) (McCouch *et al.*, 1997, Mohan *et al.*, Mohan *et al.*, 1989).

In the map-based cloning of Xa-1 (Xanthomonas campestris pv. oryzae resistance) in rice, three Xa-1 linked RFLP markers were identified, and one YAC clone with an insertion of 340kb was confirmed to possess the Xa-1 allele (Yoshimura *et al.*, 1995, Yoshimura *et al.*, 1996). The rice gene Xa-21 which confers resistance to Xanthomonas campestris pv. oryzae race 6 was already cloned by the map-based cloning (Ronald 1997, Ronald 1992, Wang 1995, Williams 1996). The transgenic plants showed high levels of resistance to pathogens (Song 1995).

Mapping of quantitative trait loci (QTLs) provide precious information for establishing breeding programs, and the extensive cooperation among various institutions is desirable for this purpose, using common materials such as recombinant inbreds, doubled haploids and backcross families under different environmental conditions (Tanksley 1993, Yano and Sasaki 1997).

Genetic linkage maps constructed on the basis of orthologous loci were compared with each other in rice, wheat and maize. It was revealed that gene content and orders are highly conserved between different species within the grass family. Some chromosome rearrangements involving inversions and transpositions have arisen during or after speciation (Devos and Gale 1997, Dunford *et al.*, 1995, Moore 1995a, Moore 1995b, Moore 1997, Van Deynze 1995). Comparative maps may provide an opportunity to identify not only major gene loci but also QTLs of agricultural importance such as disease and insect resistance, heterosis and yield (Laurie 1997, Lu 1996, Paterson 1995).

Construction of a rice physical map covered by YAC clones which have been arranged over half of the genome length was presented by the Rice Genome Research Program in Japan (. Kurata 997). Physical maps are an important resource for most molecular research facilitating positional cloning of conventional genes, sequencing of genomic DNAs and analysis of chromosome and genome structure in detail.

In situ hybridization is another effective method for localizing specific DNA or RNA sequences on chromosomes (Fukui 1996,. Fukui *et al.*, 1994,. Jiang *et al.*, 1995). Rice 5S and 17S rRNA genes (rDNA) have been mapped simultaneously by multicolor fluorescence in situ hybridization (McFISH) (Ohmido 1995).

Recently, transformation techniques have been made available for re-inserting DNA to examine gene functions and provide new breeding tools (Christou 1997,. Hiei *et al.*, 1997, Hodges *et al.*, 1991, Potrykus 1990).

Thus, genome study in rice is going to enter into a new era. By applying the basic information on new linkage mapping, many efficient and reliable means for rice breeding will be developed in near future.

The author is greatly indebted to Dr. I. Takamure from Faculty of Agriculture, Hokkaido University for his assistance.

REFERENCES

Ahn, S., J.A. Anderson, N.E. Sorrels and Tanksley, S.D. 1993. Homologous relationships of rice, wheat and maize chromosome. Mol. Gen. Genet. 241: 483-490.

Causse, M.A., T.M. Fulton, Y.G. Cho, S.N. Ahn, J. Chunwongse, K. Wu, J. Xiao, Z. Yu, P.C. Ronald, S.E. Harrington, G. Second, S.R. McCouch and Tanksley, S.D. 1994. Saturated molecular map of the rice genome based on an interspecific backcross population. Genetics 138. 1251-1274.

Chao, L.F. 1928. Linkage studies in rice. Genetics 13: 133-169.

Christou, P. 1997. Rice transformation: bombardment. Plant Mol. Biol. 35: 197-203.

Crops Research. 1963. Rice gene symbolization and linkage groups. U.S.D.A. Agr. Res. Serv. 34-28: 1-56.

Devos, K.M. and Gale, M.D. 1997. Comparative genetics in the grasses. Plant Mol. Biol. 35: 3-15.

Dunford, R.P., N. Kurata, D.A. Laurie, T.A. Money, Y. Minobe and. Moore, G. 1995. Conservation of fine-scale DNA marker order in the genomes of rice and the Triticeae. Nucl. Acids Res. 23: 2724-2728.

Fukui, K. 1996. Advances in rice chromosome research, 1990-95. Rice Genetics III. pp.117-130. IRRI, Manila, Philippines.

Fukui, K., N. Ohmido and. Khush, G.S 1994. Variability in rDNA loci in the genus Oryza detected through fluorescence in situ hybridization. Theor. Appl. Genet. 87: 893-899.

Hiei, Y., T. Komari and Kubo, T. 1997. Transformation of rice mediated by Agrobacterium tumefaciens. Plant Mol. Biol. 35: 205-218.

Hodges T.K., J. Peng, L.A. Lyznik and Koetje, D.S. 1991. Transformation and regeneration of rice protoplasts, In Rice Biotechnology (Khush, G.S. and Toenniessen, G.H. eds).. pp. 157-174. IRRI. C.A.B. International, Wallingford, UK.

Iwata, N. and Omura, T. 19971a. Linkage analysis by reciprocal translocation method in rice plants (Oryza sativa L). I. Linkage groups corresponding to the chromosome 1, 2, 3 and 4. Japan. J. Breed. 21(1): 19-28.

Iwata, N. and Omura, T. 19971b. Linkage analysis by reciprocal translocation method in rice plants (Oryza sativa L). II. Linkage groups corresponding to the chromosomes 5, 6, 8, 9, 10 and 11. Sci. Bull. Fac. Agr. Kyushu Univ. 25(3.4): 137-153.

Iwata, N. and Omura, T. 1984. Studies on the trisomics in rice plants (Oryza sativa L).. VI. An accomplishment of a trisomic series in japonica rice plants. Jpn. J. Genet. 59(3): 199-204.

Jiang, J., B.S. Gill, G.-L. Wang, P.C. Ronald and. Ward, D.C 1995. Metaphase and interphase fluorescence in situ hybridization mapping of the rice genome with bacterial artificial chromosomes. Proc. Natl. Acad. Sci. USA 92: 4487-4491.

Kadam, G.S. and Ramiah, K. 1943. Symbolization of genes in rice. Indian J. Genet. Plant. Breed. 3: 7-27.

Khush, G.S. and Kinoshita, T. 1991. Rice karyotype, marker genes, and linkage groups. In Rice Biotechnology. G.S. Khush and G.H. Toenniessen eds. pp. 83-108. IRRI. C.A.B. International, Wallingford, UK.

Khush, G.S., R.J. Singh, S.C. Sur and Librojo, A.L. 1984. Primary trisomics of rice: Origin, morphology, cytology and use in linkage mapping. Genetics 107: 141-163.

Khush, G.S., K. Singh, T. Ishii, A. Parco, N. Huang, D.S. Brar and Multani, D.S. 1996. Centromere mapping and orientation of the cytological, classical, and molecular linkage maps of rice. Rice Genetics III. pp. 57-75. IRRI, Manila, Philippines.

Kinoshita, T. 1984. Gene analysis and linkage map. In Biology of Rice. S. Tsunoda and N. Takahashi eds. pp.187-274. JSSP•^Elsevier, Tokyo.

Kinoshita, T. 1986. C. Report of the committee on gene symbolization, nomenclature and linkage groups. RGN 3: 4-5.

Kinoshita, T. 1995. C. Report of committee on gene symbolization, nomenclature and linkage groups. RGN 12: 9-153.

Kinoshita, T. 1997. 3 Gene analyses, 6. RFLP mapping. In Science of the Rice Plant: Genetics. (Matso, T. *et al.* eds). pp. 197-251, 761-781. Nobunkyo, Tokyo.

Kinoshita, T. 1997. Catalogue of gene symbols for rice:1996. In Rice Genetic Resources. pp. 622-703. National Institute of Genetics, Mishima.

Kinoshita, T. and Takahashi, M. 1991. The one hundredth report of genetical studies of rice plant. -Linkage studies and future prospects-. J. Fac. Agr. Hokkaido Univ. 65: 1-61.

Kishimoto, N., M.R. Foolad, E. Shimosaka, S. Matsuura and Saito, A. 1993. Alignment of molecular and classical linkage maps of rice, Oryza sativa. Plant Cell Reports 12: 457-461.

Kurata, N., Y. Nagamine, K. Yamamoto, Y. Harushima, N. Sue, J. Wu, B.A. Antonio, A. Shomura, T. Shimizu, S-Y. Lin, T. Inoue, A. Fukuda, T. Shimano, Y. Kuboki, T. Toyama, Y. Miyamoto, T. Kirihara, K. Hayasaka, A. Miyao, L. Monna, H.S. Zhong, Y. Tamura, Z-X. Wang, T. Monna, Y. Umehara, M. Yano, T. Sasaki and Minobe, Y. 1994a. A 300 kilobase interval genetic map of rice including 883 expressed sequences. Nature Genetics. 8: 365-370.

Kurata, N., M. Yano, Y. Minobe and Gale, M. 1994b. Conservation of genome structure between rice and wheat. Bio•^Technology 12: 276-278.

·Kurata, N., Y. Umehara, H. Tanoue and Sasaki, T. 1997. Physical mapping of the rice genome with YAC clones. Plant Mol. Biol. 35: 101-113.

Laurie, D. A. 1997. Comparative genetics of flowering time. Plant Mol. Biol. 35: 167-177.

Liu, J., Ch. Hara, M. Umeda, Y. Zhao, T.W. Okita and Uchimiya, H. 1995. Analysis of randomly isolated cDNAs from developing endosperm of rice (Oryza sativa L).: evaluation of expressed sequence tags, and expression levels of mRNAs. Plant Mol. Biol. 29: 685-689.

Lu, C., L. Shen, Z. Tan, Y. Xu, P. He, Y. Chen and. Zhu, L 1996. Comparative mapping of QTLs for agronomic traits of rice across environments using a doubled haploid population. Theor. Appl. Genet. 93: 1211-1217.

Maheswaran, M., P.K. Subudhi, S. Nandi, J.C. Xu, A. Parco, D.C. Yang, and Huang, N. 1997. Polymorphism, distribution, and segregation of AFLP markers in a doubled haploid rice population. Theor. Appl. Genet. 94: 39-45.

Matsuo, T. *et al.* eds. 1997. Science of the Rice Plant: Genetics. pp.1-1003. Nobunkyo, Tokyo.

McCouch, S.R., G. Kochert, Z.H. Yu, Z.Y. Wang, G.S. Khush, W.R. Coffman and S.D. Tanksley 1988. Molecular mapping of rice chromosomes. Theor. Appl. Genet. 76: 815-829.

36. McCouch, S.R., X. Chen, O. Panaud, S. Temnykh, Y. Xu, Y.G. Cho, N. Huang, T. Ishii and Blair, M. 1997. Microsatellite marker development, mapping and applications in rice genetics and breeding. Plant Mol. Biol. 35: 89-99.

Misro, B. 1981. Linkage studies in rice (Oryza sativa L).. 10. Identification of linkage groups in indica rice. Oryza Cuttack 18(4): 185-195.

Mohan, M., S. Nair, A. Bhagwat, T.G. Krishna, M. Yano, C.R. Bhatia and Sasaki, T. 1997. Genome mapping, molecular markers and marker-assisted selection in crop plants. Mol. Breed. 3: 87-103.

Moore, G., K.M. Devos, Z. Wang and Gale, M.D. 1995a. Grasses, line up and form a circle. Current Biology 5: 737-739.

Moore, G., T. Foote, T. Helentjaris, K. Devos, N. Kurata and Gale, M. 1995b. Was there a single ancestral cereal chromosome? TIG 11: 81-82.

Moore, G., L. Arago•Ln-Alcaide, M. Roberts, S. Reader, T. Miller and Foote, T. 1997. Are rice chromosomes components of a holocentric chromosome ancestor? Plant Mol. Biol. 35: 17-23.

Nagamura, Y., B.A. Antonio and Sasaki, T. 1997. Rice molecular genetic map using RFLPs and its applications. Plant Mol. Biol. 35: 79-87.

Nagao, S. 1935. On the gene symbolizations in rice. Agric. & Hort. 10(6) 1391-1394.

Nagao, S. 1951. Genic analysis and linkage relationship of characters in rice. Advances in Genetics 4: 181-212.

Nagao, S. and Takahashi, M. 1963. Trial construction of twelve linkage groups in Japanese rice. Genetical studies on rice plant, XXVII. J. Fac. Agr. Hokkaido Univ. 53(1): 72-130.

Ohmido, N. 1995. Rice chromosomes studies by fluorescence in situ hybridization with special reference to physical mapping and chromosome structure. J. Fac. Agr. Hokkaido Univ. 66: 277-320.

Paterson, A.H., Y.-R. Lin, Z. Li, K.F. Schertz, J.F. Doebley, S.R.M. Pinson, S.-C. Liu, J.W. Stansel and Irvine, J.E. 1995. Convergent domestication of cereal crops by independent mutations at corresponding genetic loci. Science 269: 1714-1718.

Potrykus, I. 1990. Gene transfer to cereals: an assessment. Bio•^Technology 8: 535-542

Ronald, P. C. 1997. The molecular basis of disease resistance in rice. Plant Mol. Biol. 35: 179-186.

Ronald, P.C., B. Albano, R. Tabien, L. Abenes, K. Wu, S. McCouch and Tanksley, S.D. 1992. Genetic and physical analysis of the rice bacterial blight disease resistance locus, Xa21. Mol. Gen. Genet. 236: 113-120.

Saito, A., M. Yano, N. Kishimoto, M. Nakagahra, A. Yoshimura, K. Saito, S. Kuhara, Y. Ukai, M. Kawase, T. Nagamine, S. Yoshimura, O. Ideta, R. Ohsawa, Y. Hayano, N. Iwata and Sugiura, M. 1991. Linkage map of restriction fragment length polymorphism loci in rice. Japan. J. Breed. 41: 665-670.

Sasaki, T., J. Song, Y. Koga-Ban, E. Matsui, F. Fang, H. Higo, H. Nagasaki, M. Hori, M. Miya, E. Murayama-Kayano, T. Takiguchi, A. Takasuga, T. Niki, K. Ishimaru, H. Ikeda, Y. Yamamoto, Y. Mukai, I. Ohta, N. Miyadera, I. Havukkala and Minobe, Y. 1996. Toward cataloguing all rice genes: large-scale sequencing of randomly chosen rice cDNAs from a callus cDNA library. Plant J. 6(4): 615-624.

Sato. S., T. Kinoshita and Takahashi, M. 1973. Linkage analysis of rice plant, by the use of Nishimura's recipocal-translocation lines. -Genetical studies on rice plant, LIV-. Mem. Fac. Agr. Hokkaido Univ. 8(4): 367-376.

Singh, K., D.S. Multani and Khush, G.S. 1996a. Secondary trisomics and telotrisomics of rice: Origin, characterization, and use in determining the orientation of chromosome map. Genetics 143: 57-529.

Singh, K., T. Ishii, A. Parco, N. Huang, D.S. Brar and Khush, G.S. 1996b. Centromere mapping and orientation of the molecular linkage map of rice (Oryza sativa L).. Proc. Natl. Acad. Sci. USA 93: 6163-66168.

Song, W.-Y., G.-L. Wang, L.-L. Chen, H.-S. Kim, L.-Y. Pi, T. Holsten, J. Gardner, B. Wang, W.-X. Zhai, L.-H. Zhu, C. Fauquet and Ronald, P. 1995. A receptor kinase-like protein encoded by the rice disease resistance gene, Xa21. Science 270: 1804-1806.

Takahashi, N. 1923. An example of linkage relation in rice (Preliminary report). Jpn. J. Genet. 2: 23-30.

Tanksley, S.D. 1993. Mapping polygenes. Annu. Rev. Genet. 27: 205-233.

Tanksley, S.D., N.D. Young, A.H. Paterson and Bonierbale, M.W. 1989. RFLP mapping in plant breeding: New tools for an old science. Bio•^Technology 7: 257-264.

Van Deynze, A.E., J.C. Nelson, E.S. Yglesias, S.E. Harrington, D.P. Braga, S.R. McCouch and. Sorrells, M.E 1995. Comparative mapping in grasses. Wheat relationships. Mol. Gen. Genet. 248: 744-754.

Wang, G-L., T.E. Holsten, W-Y. Song, H.-P. Wang and Ronald, P.C. 1995. Construction of a rice bacterial artificial chromosome library and identification of clones linked to the Xa-21 disease resistance locus. Plant J. 7: 525-533.

Williams, C.E., B. Wang, T.E. Holsten, J. Scambray, F. de Assis Goes da Silva and Ronald, P.C. 1996. Markers for selection of the rice Xa21 disease resistance gene. Theor. Appl. Genet. 93: 1119-1122.

Xiao, J., T. Fulton, S. McCouch, S. Tanksley, N. Kishimoto, R. Ohsawa, Y. Ukai and Saito, A. 1992. Progress in integration of the molecular maps of rice. RGN 9: 124-128.

Yamaguchi, Y. 1926. Kreuzungsuntersuchungen an Reispflanzen. I. Genetische Analyse der Granne, der Spelzenfarbe und der Endospermbeschaffenheit bei einigen Sorten des Reises. Ber. Ohara. Inst. f. Landw. Forsch. 3: 1-126.

Yamamoto, K. and Sasaki, T. 1997. Large-scale EST sequencing in rice. Plant Mol. Biol. 35: 135-144.

Yano, M. and Sasaki, T. 1997. Genetic and molecular dissection of quantitative traits in rice. Plant Mol. Biol. 35: 145-153.

Yoshimura, A., N. Iwata and Omura, T. 1982. Linkage analysis by reciprocal translocation method in rice plants (Oryza sativa). III. Marker genes located on chromosome 2, 3, 4 and 7. Japan. J. Breed. 32(4): 323-332.

Yoshimura, A., O. Ideta and Iwata, N. 1997. Linkage map of phenotype and RFLP markers in rice. Plant Mol. Biol. 35: 49-60.

Yoshimura, S., A. Yoshimura, R.J. Nelson, T.W. Mew and Iwata, N. 1995. Tagging Xa-1, the bacterical blight resistance gene in rice, by using RAPD markers. Breed. Sci. 45: 81-85.

Yoshimura, S., Y. Umehara, N. Kurata, Y. Nagamura, T. Sasaki, Y. Minobe and Iwata, N. 1996. Identification of a YAC clone carrying the Xa-1 allele, a bacterial blight resistance gene in rice. Theor. Appl. Genet. 93: 117-122.

Chapter 13

Classical Mutation Breeding and Molecular Methods for Genetic Improvement of Ornamentals

S.K. Datta & Debasis Chakrabarty

Floriculture Section, National Botanical Research Institute, Lucknow – 226 001 (U.P.), India

Floriculture has become very important industry in many countries as a result of science based techniques and steady supply of improved plant materials. World floriculture industry was estimated about US $ 4 billion in 1991 and is increasing at a rate of 8-10 per cent per year (Anonymous 1992). Cut flower is one of the major component in floriculture industry. The Netherlands plays a major role in the export of cut flowers and pot plants, being responsible for 60 per cent of world exports of cut flowers and 50 per cent of world export of pot plants. The most important commercial cut flower species are carnation, chrysanthemum, rose, gerbera, tulip, lily, freesia and daffodil. Rose, carnation and chrysanthemum dominate the sale of cut flowers which together contribute about 45 per cent of the total Dutch cut flower exports. With increasing demand of world floriculture trade, floricultural activities are increasing in developing countries and sale of floricultural products at domestic market is also increasing. Floriculture related activities are also being focussed on commercially important cut flower species. Ornamental crops include a large variety of plants. The possibilities for creating different forms and improving different ornamentals are infinite and a breeder will always have future goals to work towards. Ornamental value is the main criteria for marketing potentiality of floricultural crops. Colour and shape/size of flower and their post-harvest life are some of the major criteria in floriculture market. All the present day ornamental varieties and their novelties are the result of extensive hybridization, spontaneous and induced mutation and selection.

For developing new verities through conventional breeding breeders cross two plants with little knowledge of their genetic make up and select desirable progeny from segregating populations. Due to heterozygous nature desirable genotypes are maintained through vegetative propagation in most of the ornamentals. This method has contributed appreciable amount of new varieties in floriculture trade. A new cut flower variety is typically bred by screening many thousands of highly heterozygous progeny of sexual crosses, followed by selection of superior individuals for further trialing. The selection and propagation of highly heterozygous, superior individuals has been an extremely successful breeding scheme for cut flowers. However, difficulty of producing isogenic lines, limited gene pool, limited crossability among different genus/species, lack of knowledge regarding their genetic make up etc. have severely limited classical flower breeding. Some of the high valued floricultural crops like Chrysanthemum, rose, gladiolus, carnation etc. are highly heterozygous. Their improvement through classical breeding programme become limited. In classical breeding, one or both parents almost always possess undesirable genotype that need to be removed by extensive breeding programmes involving back crossing and recurrent selection. Inspite of such constraints new varieties are continuously being introduced through such breeding programme.

Mutation breeding is an established method for crop improvement. Induced somatic mutation breeding holds promise for effective improvement of ornamental crops. Induced mutations have a high potential for bringing about genetic improvements. Mutation techniques by using ionizing radiations and other mutagens have successfully produced quite a large number of new promising varieties in different ornamental plants by treating seed or other plant parts. Ornamental crops include a large variety of different plants including flowering and foliage pot plants, cut flower crops, bulbous annual and perennial crops as well as trees and shrubs. Attempts have been made for induction of mutations in all groups of plants. It has been well proved that the crops which are propagated vegetatively are very suitable for the application of mutation breeding methods. The main advantage of mutation induction in vegetatively propagated crops is the ability to change one or a few characters of an otherwise outstanding cultivar without altering the remaining and often unique part of the genotypes (Broertjes 1968). Mutation breeding has been most successful in ornamental plants due to some additional advantages. Changes in any phenotypic characters like colour, shape or size of flower and chlorophyll variegation in leaves can be easily detected. Many ornamental species are highly heterozygous and are often Polyploid or aneuploid and propagated vegetatively. These species offer high mutation frequency and allows the detection, selection and conservation of mutants in the M_1V_1 generation.

Voluminous literature are available in the field of application of mutagens for improvement of ornamental plants. It will not be possible to consider all the publications in the present report due to space limitations. A large number of review papers covering details of prospects and utilization of induced mutation have been compiled and published as review papers (Gustafsson 1947, 1969, Konzak 1956, 1957, Gaul 1958, 1964, 1965, Smith 1958, Swaminathan 1961, 1963, 1971, tube 1959, Gottschalk and Wolff 1983).

Important publications mentioning the list of released, commercialized and approved mutant varieties in different crop plants including ornamentals are available (Sigurbjorson and Micke 1969, Broertjes and Van Harten 1978, Maluszynski *et al.*, 1995, Anonymous 1994, Bhatia 1991, Kawai and Amano 1991, Wang 1991). Many important aspects of mutagenesis in vegetatively propagated ornamentals have been reported earlier in review article (Nybom and Koch 1965, Privalov 1967, Broertjes 1969 a, b, c, Ohba

1971, Broertjes and Van Harten 1978).

Biotechnology, the world's fastest growing and most rapidly changing technology has revolutionized research activities in the area of agriculture. Intensive research in the molecular and cellular biology of important ornamental crops have resulted in the development of effective gene transfer procedures, with subsequent recovery of engineered plants. Recent advances in molecular biology has made it possible to overcome the barriers of crossability within limited species. New varieties with aesthetic and commercial value can be produced by isolating a particular desirable gene from one line and transferring it into another line. Molecular breeding is being considered as more powerful method over classical breeding and induced mutation breeding because it can change a specific characteristic of a plant without altering other desirable characteristics in a highly directed fashion. Flower crops are marketed based on their ornamental value. The colour and form of the flower, the architecture of the plant, and the post production life of the flower are all viable targets for improvement by genetic engineering. In floriculture industry flower colour has become one of the major target area for application of genetic engineering. Carotenoid, betalains, flavonols, chalcones, aurones and anthocyanin are the main chemical components together decide the final colouration of the flower. Biosynthetic pathway of anthocyanins and other flavonoid components via the phenylpropanoid pathway has been studied using maize (Dooner 1983), Petunia (de Vlaming *et al.*, 1984) and Antirrhinum (Martin *et al.*, 1987). Extensive research resulted identification of genes encoding enzymes of this pathway.

Induced mutation breeding activities and molecular approach for development of novel ornamental varieties have been reviewed earlier by a number of workers (Mol *et al.*, 1989, Datta 1997 a, b, c, Woodson 1991, Hutchinson *et al.*, 1992, Robinson and Firoozabady 1993, James 1983, Jenks and Feldmann 1997, Schum and Preil 1998). Attempts will be made in the present review to prepare a comparative status of both classical and molecular methods applied on some common floricultural crops for their improvement. Emphasis will be given to elaborate some of the interesting findings of classical mutation breeding and prospects of molecular manipulation for floricultural crops.

Chrysanthemum

Extensive work has been done on *Chrysanthemum morifolium* Ramat. for its improvement through induced mutation by a number of workers and a wide range of physical and chemical mutagens have been used. 198 commercial mutant varieties have been reported from various countries (Datta 2000 in press). The mutant character comprises flower colour, shape and size and also physiological and most of the mutants have been developed through × or gamma irradiation.

Colchicine (0.0625 per cent) has been successfully used for development of flower colour mutation in chrysanthemum cv. Sharad Bahar. The original colour of Sharad Bahar was purple whereas the mutant colour was Terracotta Red. The mutant has been released in the name of 'Colchi Bahar' (Datta 1987, Datta and Gupta 1984, 1987). Cyotomorphological and biochemical analysis revealed that colchicine treatment can produce gene mutation (Colchi-mutation, C-mutation, the term first used by one of the author, Datta 1985a).

Recurrent irradiation method has been utilized in mutation breeding of chrysanthemum. Greater range of genetic variability (mutation frequency and spectrum) was recorded in recurrent irradiated population (Datta 1991).

Irradiation of mutant genotype resulted development of further commercial mutant varieties (Broertjes *et al.*, 1980, Datta 1985b, 1996).

Series of publication are available on standardization of regeneration protocol using different explants and *Agrobacterium* mediated Bt II, NPT II, GUS, PFG, hygromycin phosphotransferase, CHS, TSWV nucleocapsidprotein, flavanoid 3¢ 5¢ hydroxylase genes (Wordragen and Howe 1990, Wordragen *et al.*, 1992, Ledger *et al.*, 1991, Bush *et al.*, 1991, Kurle *et al.*, 1993, Jong *et al.*, 1993, Jong *et al.*, 1994, Pavingerova *et al.*, 1994, Renou *et al.*, 1993, Lowe *et al.*, 1993, Urban *et al.*, 1994, Fukai *et al.*, 1995, Burchi *et al.*, 1996, Dolgov *et al.*, 1995, Seiichi *et al.*, 1995, Benetka and Pavingerova 1995, Sherman *et al.*, 1996, Sherman *et al.*, 1998, Boase *et al.*, 1998, Kim *et al.*, 1998, Kim *et al.*, 1998, Takatsu *et al.*, 1998, Yepes *et al.*, 1999, Courtney *et al.*, 1993, 1994)).

Transgenic plants having genes of economic important characters have been regenerated from four cultivars and attempts are being made to incorporate Lc regulatory gene from maize and to produce pelargonidin-based reds through inhibition of flavonoid 3¢-hydroxylase gene activity using antisense strategy (Winefield *et al.*, 1994). Lemieux *et al.* (1990) transformed four cultivars using peduncle explants with *A. tumefaciens* and developed a white variant of 'Moneymaker' via Co-suppression of the Chalcone synthase gene. About 800 transgenic plants have been produced to date (Robinson and Firoozabady 1993).

Rose

Mutation breeding has been most successful in roses in inducing novelties and quite a large number of promising varieties have been produced which are of direct use in floriculture trade. The mutations were mostly in flower colour and shape. Both physical and chemical mutagens have been successful in inducing mutations and the mutation breeding work has been reviewed by Datta (1997d). About 30 mutants have been commercialized. Mutants with changed flower colour and/or with higher oil content or better oil quality were obtained in essential oil bearing roses by radiation and chemicals (Guo *et al.*, 1983, Klimenko and Zykov 1977, Raev 1984).

Recurrent gamma irradiation showed cumulative effects and percentage of somatic mutation and spectrum of mutation were higher after recurrent irradiation in comparision to single irradiation (Datta 1986a).

Gupta and Datta (1983) and Datta and Gupta (1985a) successfully used colchicine for the first time to induce flower colour mutation in rose cv. Contempo. It has been pointed out that normally after colchicine treatment attention is paid to chromosome duplication and its effect on phenotype. When there is no polyploid formation and when there is no gigantism in desired characters in induced polyploid in particular taxa, colchicine breeding is throught to be unsuccessful. But careful observations have led to the understanding that although colchicine is known more familiarly as a polyploidizing agent, it may also be used as a very good mutagen for flower colour change (Datta 1990a).

Neither any desirable transgenic rose nor any perfect transformation protocol have yet been developed in rose. Marchant *et al.* (1998) and Marchant *et al.* (1998) reported transgenic rose (*R. hybrida*) by biolistic method of gene delivery carrying rice chitinase gene which confer resistance to balck spot caused by the fungal pathogen *Diplocarpon rosae*. Reports are avilable on transformation protocol using different strains of *Agrobacterium* mediated NPT II, GUS genes (Short and Roberts, 1991, Robinson and Firoozabady, 1993, Noriega and Sondhal, 1991, Firoozabady *et al.*, 1991, Firoozabady *et al.*, 1994, Matsuda *et al.*, 1996, Marchant *et al.*, 1996). But most of the transformation protocols are cultivar specific and the frequency of transformation is often low.

Carnation

Carnation is considered to be one of the few plants included first in mutation breeding programme and the first mutant reported by Richter and Singleton (1955) was changed flower colour and type. More than 13 new varieties with changed flower colour, developed through X-rays and EMS treatment, have been commercialized (c.f. Broettjes and Van Harten 1988).

Agrobacterium mediated first transgenic plant of carnation was developed in 1989 (Woodson and Goldsbrough). Transgenic carnations carrying EFE and ACC synthase genes have been developed with increased vase life through reduction in senescence (Michael *et al.*, 1993). Regeneration and transformation systems have been successfully developed by a number of workers, using different explants of various carnation cultivars (Firoozabady *et al.*, 1991, Lu *et al.*, 1991, Altvorst *et al.*, 1995, Messeguer and Mele 1994, Vain Stein *et al.*, 1993, Jeffersen 1987). Recently Ovadis *et al.* (1999) developed transgenic carnation plants for selection of novel traits using rolC gene from *Agrobacterium rhizogenesis*. rolC-transgenic plants showed increased axillary bud breakage, delayed flowering, higher vegetative shoot length, higher yield, better rooting etc.

Gladiolus

Both physical (gamma rays, X-rays, fast neutrons, thermal neutrons and electric shock) and chemical (aluminium chloride, colchicine, diethylsulphate, formalin, glycol, MMS, DES, EMS, Nitroso dimethyl urea, N-nitroso-N-ethyl urethene and N-nitroso-N–methyl urethene) mutagens have been successfully used for induction of mutations through both corms and seed treatment. Mutants with improved flower form and colour, leaf colour, increased production etc. have already been marketed (Dryagina 1975, Broertjes and Bakker 1984). A good amount of interesting informations regarding mutation frequency, chimeric nature and spectrum of mutation in M_1V_1, M_1V_2 and M_1V_3 have been reported (Buiatti and Tesi 1968, Buiatti *et al.*, 1965, Buiatti *et al.*, 1967, Banerji *et al.*, 1981, Banerji and Datta 1987, Banerji *et al.*, 1994).

No gladiolus variety has yet been developed through biotechnology. Only *Agrobacterium*–mediated transformation system has been tested using corms (Graves and Goldman, 1987).

Bougainvillea

Bougainvillea is a hardy perennial, most showy ornamental plant. The suitable dose of gamma rays for irradiation of stem cuttings of single bracted bougainvillea has been standardized from 250-1250 rads. The most promising and beautiful chlorophyll variegated gamma ray induced mutants are 'Arjuna' (Gupta and Shukla 1974, Gupta and Nath 1977); 'Jayalaxmi Variegata' and 'Lady Hudson of Ceylon Variegata' and two ornamental novelty mutants are 'Jaya' and 'Silver top' (Abraham and Desai 1977). Mutation breeding is the only method for improvement of double bracted bougainvillea as improvement by conventional cross-breeding is not possible due to absence of flower. Three gamma ray induced chlorophyll variegated mutants 'Pallavi', 'Mahara Variegata' and 'Los Banos Variegata' have already been commercialized (Banerji *et al.*, 1987a, Datta and Banerji 1994, 1997, Datta 1992a). No reports are available on induced bract colour mutation.

No work has yet been initiated anywhere to induce novelty in bougainvillea using molecular technique, nor even there are any reports of *Agrobacterium*-mediated and / or biolestic transformation protocol.

Hibiscus

Semi-acute gamma rays have developed flower colour and form mutations in Hibiscus (*H. rosa sinensis*) cvs. Alipur Beauty and Cruenthus (Das *et al.*, 1974 and Das *et al.*, 1977). 'Shirasagi-no Yume' a new induced mutant variety has been commercialized in Japan (Maluszynski *et al.*, 1992). 4 krad gamma ray treated five petalled single flower type mutant 'Anjali' has been developed from double flower type cv. 'Alipur Beauty' and commercialized (Banerji and Datta 1986a, b). No reports on molecular manipulation is available

Portulaca is a succulent herb which grows in summer season. Seven gamma ray induced flower colour and flower type mutants have been developed and released–'Karna Phul' (Desai 1973, 1974) and 'Jhumka', 'Karnapali', 'Lalita', 'Mukta', 'Ratnam' and 'Vibhuti' (Gupta 1966). Gamma ray and EMS and their combined treatment induced flower colour mutation in *P. granadiflora* (Raghuvanshi and Singh 1979, 1980). No work has been initiated for development of transgenic variety.

Tuberose

Gamma rays (2 krad) have developed two chlorophyll variegated mutants and commercialized: 'Rajat Rekha' (leaves having silvery white streaks along the middle of the blade, induced in single flowered tuberose) and 'Svarana Rekha' (leaves having golden yellow streaks along the margin, induced in double flowered tuberose) (Gupta *et al.*, 1974). Gamma ray and neutron induced chimeric mutations could not be established in pure form (Younis and Borham 1975, Abraham and Desai 1976). No flower colour mutation could be induced inspite of extensive mutation breeding work, molecular approach for developing new flower colour is most essential for this important cut flower species.

Lily

Flower colour and low-light tolerant mutants have been developed thorugh X-ray treatment and commercialized (Van Groenestijn and Van Tuyl 1983). De Mol (1936 c.f. Broertjes and Van Harten 1988) released first commecial flower colour mutant "Faradya' which was developed thorugh irradiation of cv. ' Fantasy'. Eight X-ray induced mutants (with changed flower colour, flower colour + short flower stalk, variegated leaves etc). have been commercialized (De Mol. 1949, Custers *et al.*, 1977, Van Eijk and Eikelboom, 1981 a, b). Extensive work resulted accumulation of important basic informations (bulb stage, time of irradiation, selection procedure, optimum dose etc). for successful application of mutation breeding.

No desirable commercially important transgenic lily has yet been developed. Only preliminarry reports are available on testing the capability of *Agrobacterium* of transforming lily (Cohen and Meridith 1992, Langveld *et al.*, 1994).

15 cultivars with new flower colours have been induced through mutation in Rhododendron simsi (Heursel 1980; Maluszynski *et al.*, 1992).

Tulip

Series of flower colour sports have been developed in tulip due to high heterozyosity (De Mol 1933, 1949). Extensive mutation breeding activities have been done on tulip and valuable basic informations have been generated on mutagenesis of bulbous plants (Grabowska and Mynett 1970, Matsubara *et al.*, 1965, Nezu 1987, Custers *et al.*, 1977). Eight X-ray induced mutants have been commecialized (Broertjes and Van Harten 1988). Attempts are being made for developing transformation system for tulip using different strains of *Agrobacterium* and particle bombardment (Wilmink *et al.*, 1992, 1994).

Torenia (*Torenia fournieri* -Scrophulariaceae) is an important bedding plant. Suntory Ltd. Japan developed a new type of Torenia cv. Summerwave (*T. hybrida*) which had only one flower colour (Blue, SWB). Tsuda *et al.* (1999) developed series of transgenic plants with wide range of flower colour (Pure white, pale yellow, purple pink, pale blue etc). harbaring sense cDNAs encoded CHS, DFR and F35H enzymes isolated from torenia SWB.

Pelargonium

Most of the commercial varieties are seedling selections via cross breeding although reports are available for commercialized spontaneaous mutants. Mutation breeding using X-irradiation induced flower colour and leaf colour and shape mutations have been reported but no mutant variety has been commecialized (Craig 1963). Somaclonal variants are being grown commecially (Skirvin and Janick 1974, 1976). No transgenic variety has reached upto commercial level. *A. tumefaciens* inoculated explants from cotyledons and primary leaves produced.kanamycin -resistant calluses that were GUS positive (Janssen and Gardner 1989). Transformation and regeneration system for cv. Dubonnet have been developed, but transgene expression in 265 plants indicated a significant loss of gene expression (Deroles *et al.*, 1997). Stem and petiole fragments of a scented Pelargonium species was inoculated with three agropine-type *Agrobacterium rhizogens* strains. Regenerated shoots possessed improved ornamental characters and increased production of essential oil (Pellegrineschi *et al.*, 1994).

Gerbera

Considering its slow traditional method of vegetative propagation tissue culture micropropagation technique has been developed for large scale production to meet the commerial demand (Murashige *et al.*, 1974; Pierik *et al.*, 1975; Pierik *et al.*, 1982; Pierik 1991). Inspite of concentrated efforts no suitable transformation protocol has yet been reported in gerbera. Successful transformation (GUS and NPT II gene) has been confirmed by Hutchinson *et al.* (1992) but the percentage of transformation is very low. Haploid plants have been produced from callus derived from ovules (Sitbon 1981, Meynet and Sibi 1984).

Mahna and Garg (1989) induced two mutants with dentate/ dissected flower corrolas in *Petunia nyctaginiflora* treating seeds with chemical mutagens. Petunia sp. and *Antirrhinum majus* have been extensively studied for understanding the biochemistry and flavonoid biosyntheis genetics and the studies have enabled the isolation and cloning of some of the genes associated with flavonoid metabolism (Wiering and Vlaming 1984; Coen *et al.*, 1988; Van Tunen and Mol 1991). Flower colour modifications in Petunia was the first successful genetic transfromation of AI gene from *Zea mays* (Meyer *et al.*, 1987).

From the recent reviews it is difficult to estimate the exact contribution of classical mutation breeding in development of new ornamental varieties. The actual number is more than the numbers reported in different international journals and reviews. Commercial growers do not disclose details about their varieties. In addition, huge amount of literature published by several workers in their regional journals and local subject oriented literture are not accessable to international forum. These are some of the major drawbacks to estimate the quantum of mutation breeding work done on ornamentals. In most of the reviews the same example is cited to show the potentiality of mutation breeding. The quantum of contribution through mutation breeding and its exact potentiality can be properly judged if regional literature are also reviewed. This will also enrich the basic informations rleated to mutagenesis (Datta 1980, 1986b, 1990b, 1992b, Datta and Gupta 1981a, b, 1983a, b, 1985b).

Mutation breeding work done on different ornamental crops around the world have been reviewed from time to time. It is worthwhile to mention that National Botanical Research Institute (NBRI), Lucknow, India is one of the pioneer instituions where a commendable work have been done on various ornamentals. For over three decades, scientists at NBRI are engaged in both basic and applied research for the improvement of different medicinal and aeromatic plants, oil bearing plants, pulse crops in general and ornamental crops in particular. The main aim of research activities have been focussed to search out novel genes of commecial importance through induced mutagenesis. Ornamental crops included in the study were Bougainvillea, Canna, Chrysanthemum, Dahlia, Hibiscus, Gladiolus, Grusenteplitz, *Perennial Portulaca*, *Polianthes tuberosa*, *Vinca rosea*, Rose, *Lantana depressa* etc (Datta 1997a, b, c, 1998; Datta and Gupta 1985). Some of the results have already been mentioned above.

A good amount of promising mutants have been developed through in vitro mutagenesis. A number of mutant/variants with changed flower colour and morphology have been induced by irradiation of in -vitro shoots of Gerbera (Laneri *et al.*, 1990). However, official registration report of only one mutant variety is available (Jerzy and Zalewska 1996). Simard *et al.* (1992) developed a number of mutants in carnation by x-irradiation of in vitro petal culture. X-rays treatment produced flower colour variants from node culture of carnation (Cassells *et al.*, 1993). Walther and Sauer (1986, 1991) developed a series of flower colour variants through x-ray treatment of *in vitro* shoots from a heterozygous genotype of Gerbera. About 19 variants including flower colour and shape have been induced through irradiation of *in vitro* shoot culture of a pink cultivar of Gerbera. Jerzy (1990) induced tubular floret mutants through in vitro × and gamma radiation in chrysanthemum. Solid flower colour mutants were obtained after × or gamma ray treatment of in vitro leaves of chrysanthemum cv. Richmond (Jerzy 1990, Jerzy and Zalewska 1996). Nagatomi *et al.* (1996) treated shoots, developed in vitro from petals and leaf blades explants, with chronic irradiation and developed changed flower shape (funnel or bell-shaped) mutants in *Eustoma grandiflorum*.

The term 'Somaclonal Variation' was introduced to describe the genetic variation in plants regenerated from tissue culture via callus (Larkin and Scowcroft 1981) and which is considered as a novel source of variability for breeders. Although various groups have carried out research on somaclonal variation, but it is clear that the use of somaclonal variation in breeding is limited, and also in most application of biotechnology somaclonal variation is very undesirable.

In chrysanthemum, Bush *et al.* (1976), Khalid *et al.* (1989), Malure *et al.* (1991a, b) reported somaclonal variants with possible commercial value. In *Bigonia*, plants with leaf variegation have been reported by Bouman and de Klerk (1997) as a result of somaclonal variation. Griesbach (1989) described a dwarf mutant for daylili (*Hemerocallis*), obtained through tissue culture. Somaclonal variation in the form of change in flower colour, variegated leaves, floral and leaf morphology have also been reported in *Dianthus caryophyllus* (Hackett and Anderson 1967), *Lilium* (Stimart *et al.*, 1980), *Pelargonium* sp. (Skirvin and Janick 1976), *Gerbera jamesonii* (Stibon 1981) etc.

Although somaclonal variation has been extensively exploited in many species to create variants but still it has not yet resulted in the release of any new cultivar that has had a major impact on the industry.

The main bottlenecks in mutation breeding is the formation of chimeras in multicellular organism. To avoid such undesirable situation it is most essential to develop methods by which shoots or plants are obtained which orginate from only one cell. In vitro, this could be done via nono-cell culture, tissue culture and subsequent regenertion of plants. But presently it is limited to a number of species only. In vivo, however, a number of species are known in which the formation of adventitious buds can be stimulated, either on roots (phlox), bulbs (Hyacinthus) or leaves (Begonia, Saintpaulia and Streptocarpus) and bulb scales (Lilium) (Broertjes *et al.*, 1968, Haccius and Hausner 1974, Naylor and Johnson 1973, Sparrow *et al.*, 1980, Broertjes 1968).

The size of mutant sector varies from a narrow streak on a petal to entire petal to whole flower and a portion of a branch to entire branch. When the entire branch is mutated, mutants can be isolated through conventional techniques while small sectorial mutation in the floret/petal cannot be isolated using existing conventioanl propagation techniques. The difficulty in regenerating whole plants from sectorial mutated tissue is the main bottleneck and a huge number of such spontaneous and/or induced new flower colour mutants are lost every year. In vitro regeneration methods for chrysanthemum are well established (Hill 1969, Ben-Jacov and Langhans 1972, Earle and Langhans 1974a, b, Kaul *et al.*, 1990, Lu *et al.*, 1990). Reports of adventitious shoot regeneration from floret explants of chrysanthemum are also available, but in all cases shoots were produced from floret derived callus (Ahloowalia 1992, Bush *et al.*, 1976, De Jong and Custers 1986, Nikaido and Onozawa 1989, Malaure *et al.*, 1991a, b, Nagatomi *et al.*, 1993) and there is always a loss of some genetic homogenecity with a lengthy callus phase (Bush *et al.*, 1976, Sutter and Langhans 1981, Malaure 1991). Recently an efficient technique has been standardized at NBRI, Lucknow, India for direct shoot regeneration from individual florets of chrysanthemum. Using this direct shoot regeneration protocol a number of new flower colour/shape mutants have been isolated through management of induced and spontaneous mutant chimeric tissues (Charkabarty *et al.*, 1999, 2000, Mandal *et al.*, 2000a, b). Somatic mutations in flower colour were detected as chimera in gamma-ray treated populations of chrysanthemum cv. Maghi. The original floret colour was mauve whereas the mutant colours were white and yellow. Both flower colour mutations were detected after treatment with 1.5 and 2.0 krad gamma rays but in all cases, mutations were restricted to a few florets only. Direct shoot regeneration technique has been standardized and two new flower colour varieties *viz.* White and Yellow have been established in pure form (Chakrabarty *et al.*, 1999). Few florets with yellow colour were developdue to spontaneous mutation in 'Kasturba Gandhi' a large white chrysanthemum which has good demand in floriculture trade due to its beautiful shape and size. New variety with yellow floret colour has been established in pure form using chimera management technique (Chakrabarty *et al.*, 2000). Rooted cuttings of white flowered cv. 'Purnima' (mutant of 'Otome Zakura' with mauve flower) and yellowish red flowered cv. 'Colchi Bahar' (colchicine induced mutant of 'Sharad Bahar' with mauve flower) of Chrysanthemum were treated with gamma rays and sectorial somatic mutations in flower colour were detected in both the cvs. Mutated ray florets (yellow colour for both the cvs). have been established as yellow mutants through in-vitro direct shoot regeneration technique (Mandal *et al.*, 2000a).

From the literature survey it is very clear that almost all the colours are mutable and the spectrum of mutation is very high in some starting colour. It has already been tested by repeated experiments that pink genotypes of chrysanthemum are very sensitive for mutation. Series of development of new flower colours sports have been reported from single genotype in Dianthus (cv. Pallas) and chrysanthemum (cv. Horim) (Silvy and Mitteau 1986, Broertjes *et al.*, 1980). Nine flower colour/shape mutants have been developed by gamma radiation in chrysanthemum cv. Undaunted (large flowered), six flower colour mutants form one small flowered decorative pompon type chrysanthemum cv. E-13, five colour mutants from samll flowered double korean cv. Flirt, five flower colour/shape mutants from decorative type cv. D-5, five flower colour mutants form late blooming small flowered Pompon cv. Maghi and five colour mutants form rose cv. Contempo. All these mutant varieties have been commercialized (Datta 1988a, b, 1997b, c, Gupta and Datta 1990).

Radiation induced phenotypic variation including several interesting changes in flower form for novelties have been reported above and in earlier reviews. 'Nirmod' is a white single Korean type chrysanthemum in which the flower-heads are flat and disc conspicuosly visible. The mutant (Cosmonaut) developed from it had Anemone type flower head where disc florets developed prominently giving it hemispherical shape (Datta 1984). Chrysanthemum cv. D-5 is a double korean with mangolia purple flower head. Its mutant (Shabnam) developed a small appendage like structure at the tip of each floret. The appendage look like dew drops and it looks very attractive (Datta 1990c). Venkatchalam and Jayabalan (1994a, b) developed a number of flower colour mutants associated with changed floret numbers by gamma irradiation in Zinnia.

Induction of tubular florets is one of the interesting observtions in chrysanthemum. M-24 is a large flowered, spoon type purple cultivar. 1.5 krad gamma rays have changed the ray florets into tubular shape with spatula like open tips. In the mutant ('Tulika') the length of the tubular portion of the floret was more and open tip portion became reduced to give a brush like appearance of the florets (Datta *et al.*, 1985)

Development of striped flower colour mutants in some ornamentals have increased the commercial potentiality. 'Batik' is a gamma ray induced mutant with yellow stripe on red background of florets developed from chrysanthemum cv. Flirt with double korean red purple florets. One white flower colour mutant (Maghi White) has been developed in one late blooming small flowered pompon type cv. Maghi with mauve flower. 'Maghi White' after further gamma irradiation developed striped mutant (Maghi stripe) with yellow stripe on white background (Shukla and Datta 1993). Two striped mutants of rose ('Abhisarika' and 'Striped Christian Dior') have been developed through gamma irradiation and released by Indian Agricultural Research Institue, IARI, New Delhi (Kaicker 1985). 'Twinkle' (Pink stripe on cherry red background) developed form 'Imperator' with cherry red flower after gamma irradiation. 'Contempo Stripe' (yellow stripe on orange background) and 'Mrinalini Stripe' (white stripe on pink background) developed from rose cvs. 'Contempo' (orange petal with yellow eye) and 'Mrinalini' (pink) after treatment with gamma rays (Gupta *et al.*, 1992). All the above mentioned striped mutants have been commecialized and have great demand in the market both as cut flower and as pot plants.

It is commonly accepted that domestication of wild species has been conditioned by mutation following selection. Mutation experiments with wild species help to understand this domestication process in floriculture. *Lantana depressa* Naud (Vebenaceae) is a semi-wild herb with creeping habit and yellow flower with little genetic variability. It is grown on road sides and boarders of gardens. Gamma rays have been successfully induced three mutants in *L. depressa i.e. L. depressa* variegata (with variegated leaves), 'Niharika' (with mustard yellow flower) and *L. depressa* bicoloured (flower with both yellow and white) which can be grown now as potted ornamentals (Datta 1995a). Use of mutation breeding as an effective technique for the domestication of wild ornamentals has been suggested.

Mutation is a chance process–we may get or not. When any variety is irradiated–we do not know what colour change we are going to achieve. Spectrum of induced flower colour mutations was one of the very interesting findings in mutation breeding. More than one flower colour mutation has been observed in different cultivars. The work has inspired the irradiation of different cultivars for colour. From the extensive work at NBRI, India on different cultivars of chrysanthemum and rose and other ornamentals it has been precisely determined that the flower colour (pigment composition) of parent cultivars is an important indicator for new flower colour mutation. From the repeat experiments with the same and/or different cultivars following conclusions have been drawn for induction of mutation:

1. If white flower varieties are irradiated, and if there is any mutation, the mutation will either be in the flower shape or colour (yellow)
2. Red flower varieties, on the otherhand, will produce either a completely yellow mutation or a mixture of red and yellow mutation.
3. If yellow varieties are irradiated, and if there is any mutation, the mutation will be either different shades of yellow or white or mixture of yellow and white.

Work on the preparation of a colour chart based on the colour (pigment composition) of the parent cultivars and the spectrum of induced flower colour mutations is under way. From this colour chart it will be possible to decide beforehand the colour combination of mutant expected to develop from a particular colour variety or breeder will be able to create the desired flower colour mutation (directive mutation) when necessary for the floriculture trade by selecting specific varieties with specific colour (Datta 1995b).

For genetic engineering to be effective in delivering new plants for floriculture, three most important componenets are needed.

1. A suitable regeneration system.
2. An efficient transformation system.
3. Suitable genes that can be cloned and can impart favourable traits.

There is substantial work on regeneration of various ornamental species including Carnation (Miller *et al.*, 1991, Miller *et al.*, 1991, Nugent *et al.*, 1991; Firoozabady 1991, Simard *et al.*, 1992), chrysanthemum (Kaul *et al.*, 1990; Lu *et al.*, 1990; May and Trigiano *et al.*, 1991, Urban *et al.*, 1994, Chakrabarty *et al.*, 1999), rose (Hill 1967, Rout *et al.*, 1992, Marchant *et al.*, 1996, Kintzios 1999), gerbera (Pierik *et al.*, 1975, Murashige *et al.*, 1974, Jerzy and Lunomski 1991), tulip (Orlikowska *et al.*, 1999, Reynoird *et al.*, 1993, Ruffonie and Bmassabo 1991) and lily (Wilmink *et al.*, 1992).

There have been several reports on the transformation of ornamentals and most of the ornamental crop that have been tranformed rely on *Agrobacterium* mediated transformation (Babu *and Chawala* 2000, Wordragen and Howe 1990, Wordragen *et al.*, 1992, Ledger *et al.*, 1991, Bush *et al.*, 1991, Kurle *et al.*, 1993, Jong *et al.*, 1993, 1994, Renou *et al.*, 1993, Lowe *et al.*, 1993, Urban *et al.*, 1994, Fukai *et al.*, 1995, Burchi *et al.*, 1996, Dolgov *et al.*, 1995, Seiichi *et al.*, 1995, Benetka and Pavingerova 1995, Sherman *et al.*, 1996, Sherman *et al.*, 1998, Boase *et al.*, 1998, Kim *et al.*, 1998; Kim *et al.*, 1998, Takatsu *et al.*, 1998, Yepes *et al.*, 1999 Firoozabady *et al.*, 1991, Lu *et al.*, 1991, Altvorst *et al.*, 1995, Messeguer and Mele 1994, Vein Stein *et al.*, 1993, Jeffersn 1987).

Recent progress on molecular biology allows to characterize and clonning of several important genes. Among which are, flower colour, post harvest attributes, flower shape, disease resistance, regulation of flowering, abiotic stress etc.

One of the major target of classical breeder, mutation breeder and molecular breeder is development of value added desirable new and novel flower colour. Presently available colourful varieties in the trade are the results of successful contribution of classical breeding and induced/spontaneous mutation breeding. Identification of genes involved in the pathway of biosyntheis of pigments (flavonoids) in maize, petunia and Antirrhium has opened a new way for developing new flower colour through gene manipulations (Dooner 1983, Wiering and de Vlaming 1984, de Vlaming *et al.*, 1984, Matin *et al.*, 1987, Coen *et al.*, 1988; Van Tunen and Mol 1991). The most exciting era of pigment manipulation started since the first successful transformation of maize A1 gene in Petunia (Meyer *et al.*, 1987). Genetically engineer flower colour transgenic petunia plants have been developed using antisense technology (Van der Krol *et al.*, 1988, Koes *et al.*, 1986). Some of the major ornamental crops like Carnation, Chrysanthemum, Gerbera and rose do not occur hydroxylation enzyme (3, 5-hydroxylase) responsible for blue colour. Exciting work is on progress for production of blue flowers (Holton *et al.*, 1992).

Ornamental species is threatened by a wide range of disease and pests. Development of disease resistance varieties was not in the priority list of conventional breeding due to an extensive and successful application of chemical control. Recently much attention has been paid for developing disease resistance strain both by conventional cross breeding, induced mutation breeding and molecular breeding to avoid environmental pollution due to chemical control.

Huge amount of flower crops are lost due to virus infection. Genetically engineered virus resistance technology will be an exciting area for protection of ornamental crops against viral infection (Abel *et al.*, 1986, Hoekema *et al.*, 1989, Tumer *et al.*, 1987). An effective molecular method has been developed for protection against the infecting virus through expression of a viral coat protein gene in transgenic plants (Powell-Abel *et al.*, 1986, Buck 1991). Reports of such molecular approach to control the major virus disease in ornamental crops are scanty except some preliminary work on carnation (c.f. Hutchinson 1992).

Cytokinin has been found to be very effective in improving vase life through molecular control of flower senescence (Halevy and Mayak 1981). Ethylene biosynthetic pathway have been properly documented and important genes for ACC synthase, EFF or ACC oxidase have been cloned and sequenced (Kende 1989, Nakajima *et al.*, 1990, Van der Straeten *et al.*, 1990). Carnation petals do not produce autocatalytic ethylene and do not senescence rapidly when treated with BAP and exposed to ethylene (Mor *et al.*, 1983). Such molecular control of flower senescence will create a great impact to extend storage and vase life in cut flower industry (Lawton *et al.*, 1989, Woodson *et al.*, 1990, Wang and Woodson 1991). Interesting observations have been reported on the role of GA in a wide range of development process including petal growth and pigmentation in petunia flowers (Leitner–Dagan *et al.*, 1999)

One of the recent exciting era of genetic engineering is manipulation of flower shape since the systematic analysis of floral homeotic mutants were done on *Arabidopsis* (Bowman *et al.*, 1989). Deficiens genes and agamous gene, isolated form *Anthirrhinum majus* (Sommer *et al.*, 1990) and *A. thaliana* (Yanofsky *et al.*, 1990) respectively, have created interest to increase the commercial value of cut flowers with novel flower shapes through molecular manipulation of floal homeotic gene. Recently, Griesbach (2000) after suceessful testing on gladiolus and petunia recommended faster gene gun genetic technique for developing new colours in orchids. He bombarded the orchid petals with microscopic gold coated with

DNA taken from corn plants and detected new colour within 48 hours. Griesbach said the technique could be a boon to breeders of orchids because of the flowers long generation time up to six years in some commercial types.

Despite the great progress and interest in genetic transformation in these crops, there is no any transgenic commecial cultivars having economic important genes. However, the first genetically engineered floral crop having economically desirable genes were reported in petunia (Mayer *et al.*, 1987) and chrysanthemum (Courtney-Gutterson *et al.*, 1993, Courtney-Gutterson 1994), both with altered flower colour. There is no report of virus resistant floral crops developed through genetic engineering. Recently two cultivars of chrysanthemum have been engineered with the N-gene of TSWV and recovered TSWV resistant plants (Sherman *et al.*, 1993, 1996).

If the rate of technological advancement and implification of recombinant methodologies continues at its current pace, in the near future then will be so many commercial transgenic ornamental cultivars sold in the market.

From the present review it is very clear that majority of commercial varieties of ornamental species have been developed through conventional breeding, induced mutation and selection. Every technique has its own adventages and disadvantages. Mutation breeding at its present status appears to be well standerdized, efficient and cost effective. Although mutation breeding is a random (chance) proces, reports are available for directive mutation with some starting colour. Molecular technique is expected to be as per desire but it will take more time to develop a reproducible regeneration and transformation system. Except few, all induced new flower colour mutants are stable. But there are numerous reports for the variation and instability of genes in transgenic plants. It has been reported that a gene can loose its transcriptional activity after crosses into the next generation or after the introduction of another foreign construct into such a transgenic line (Matzke *et al.*, 1989).

Recent promising and exciting results, obtained in floriculture crops due to considerable effort to isolate and understand the structure, function and regulation of genes of commercial importance, clearly indicate that the future scope of improving ornamentals using molecular breeding are unlimited. Due to some valid restrictions of traditional breeding, breeding through biotehnology is a necessity. Molecular breeding offers new and exciting challanges for the future. It has just enabled researchers to start unreaveling the molecular details of genes of commercial novelities. Inspite of continued creation of novelities, research has declined due to limited funding projects on classical breeding and mutation breeding. Majority of the projects related to molecular biology are getting priority for financial assistance. As a result mutation breeding activities have been restricted to limited institution/universities and development of mutant varieties hasd declined. At this stage it is possible to increase the rate of mutant development by combining the classical mutation breeding and moleculart technique. This concept has been clearly proved by the recent report for management of chimera through direct regeneration from petals of chrysanthemum (Chakrabarty *et al.*, 1999, 2000, Mandal *et al.*, 2000).

Now all chimeric mutant tissue can be established in pure form which otherwise could not be isolated. Combined technique can revolutionize in floriculture. It is also most essential that all the existing ornamental varieties including mutant varieties should be properly characterized and documented. This will help not only to incorporate them for further improvement programme through traditional methods but also these will be the source materials for selection of desirable genes by molecular breeders for developing desired improved variety as per demand of floriculture trade.

REFERENCES

Abel, P.P., Nelson, R.S., De D., Hoffmann, N., Rogers, S.G., Fraley, R.T. and Beachy, R.N. 1986. Delay of disease development in transgenic plants that express the tobacco mosaic virus coal protein gene. Science, 232: 738-743.

Abraham, V. and Desai, B.M. 1976. Radiation–induced mutants in tuberose. Indian J. Geneti. Plant Breed., 36(6): 328-331.

Abraham, V. and Desai, B.M. 1977. Radiation induced variegation mutants in Bougainville. Curr. Sci., 46(10): 351-352.

Ahloowalia, B.S. 1992. In vitro radiation induced mutants in chrysanthemum. Mutation Breeding Newsletter, 39: 6.

Altvorst, A.C., Risken, T., Koehort, H. and Dons, H.J.M. 1995. Transgenic carnations obtained by *Agrobacterium tumefaciens*–

mediated transformation of leaf explants. Transgenic Research: 4: 105-113.

Anonymous 1994 List of new mutant cultivars. Mutation Breeding Newsletter. 39: 14-33.

Anonymous 1992. Review of the AIPH 1991 statisics. The Internationla Floriculture Quarterly Report 3, 32-52.

Babu, P. and Chawla, H.S. 2000. In vitro gegeneration and Aq. meidated transformation in gladiolus. J. Hort. Sci. and Biotech., 74: 400-404.

Banerji, B.K. and Datta, S.K. 1986. Induction of single flower mutant in Hibiscus cv. 'Alipur Beauty'. J. Nuclear Agric. Biol. 15(4): 237-240.

Banerji, B.K. and Datta, S.K. 1986a. Induction of single flower mutant in Hibiscus cv. Alipur Beauty. J. Nucl. Agric. Biol., 15: 237-240.

Banerji, B.K. and Datta, S.K. 1986b. 'Anjali'–A new gamma ray induced single flower mutant of Hibiscus.. J. Nuclear Agric. Biol. 17(2): 113-114.

Banerji, B.K. and Datta, S.K. 1987. Physical mutagenesis in Gladiolus–A review. North American Gladiolus Council Bulletin 87(170): 90-96.

Banerji, B.K., Datta, S.K. and Sharma, S.C. 1994. Gamma irradiation studies on Gladiolus cv. 1981. Effects of gamma irradiation on Gladiolus L. II. Cytomorphological studies on *Gladiolus psittacinus* var, Hookeri cv. Orange. NAGC Bulletin, 148: 50-57.

Banerji, B.K., Gupta, M.N. and Datta, S.K. 1981. Effects of gamma irradiation on Gladiolus L. II. Cytomorphological studies on Gladiolus psittacinus var, Hookeri cv. Orange. NAGC Bulletin, 148: 50-57.

Banerji, B.K., Nath, P. and Datta, S.K. 1987a. Mutation breeding in double bracted Bougainvillea cv. "Roseville's Delight'. J. Nuclear Agric. Biol. 16(1): 48-50.

Ben-Jaacov, J.and Langhans, R.W. 1972 Rapid multiplication of chrysanthemum plants by stem tip proliferation.HortScience 7: 289-290

Benetka.V and Pavingerova, D 1995. Phenotypic difference in transgenic plants of Chrysanthemum.Plant Breeding. 114:169-173.

Bhatia, C.R. 1991 Economic impact of mutant varieties in India I: Plant mutation breeding for crop improvement pp. 33-45. Vienna. IAEA.

Boase, M.R., Bradley, J.M and Borst, N.K. 1998. Genetic transformation mediated *by A. tumefaciens* of florists chrysanthemum *(Dendrathema grandiflora*) cultivar Peach Margaret. In vitro Cellular & Dev. Biology Plant, 334: 46-51.

Bouman, H. and de Klerk, G.J. 1997. Somaclonal variation. In: Biotechnology of ornamental plants., pp. 165-183 9eds. R.L. Geneve, J.E. Preece and S.A. Merkle). Walling ford: C.A.B. International.

Bowman, J.L., Smyth, D.R. and Meyerowitt, E.M. 1989. Genes directing flower development in Arabidopsis. Plant Cell, 3: 749-758.

Broertjes, C. 1968. Mutation breeding of vegetatively propagated crops. Fifth Congress of the European Association for Research on plant breeding. Milano, 30-9 to 2-10, 1968: 139-165.

Broertjes, C. 1969a. Mutation breeding of Streptocarpus., Euphytica: 18: 33-339.

Broertjes, C. 1969b Mutation breeding of vegetatively propagated crops: In: G.C. Chisci and G. Hanssmann (editors), Proc. 5th Eucarpia Congress, Millan 1968, gent. Agrar, 23: 139-165.

Broertjes, C. 1969c Mutation breeding of vegetatively propagated crops: In: G.C. Chisci and G. Hanssmann (editors), Proc. 5th Eucarpia Congress, Millan 1968, gent. Agrar, 23: 139-165.

Broertjes, C. and A.M., Van Harten 1978. Application of Mutation breeding methods in the improvement of vegetatively propagated crops. (Elsevier Scientific Publishing Company, Amsteerdam)

Broertjes, C. and Bakker, A.G. 1984. Mutatieveredeling van gladiolen. Bloembollen culture, 95(25): 566-567.

Broertjes, C. and Van Harten, A. M. 1988. Application of mutation breeding methods in the improvement of vegetatively propagated crops. An Interpretive literature review. Elsevier Scientific Publishing Co. Amsterdam-Oxford, New York.

Broertjes, C., Koene, P. and Van Veen, J.W.H. 1980. A mutant of a mutant irradiation of progressive radiation–induced mutants in a mutation breeding programme with *Chrysanthemum morifolium* Ramat. Euphytica, 32(10: 97-101.

Broertjes, C., Roest, S. and Bokelmann, G.S. 1976. Mutation Breeding of *Chrysanthemum morifolium* Ram. using in vivo and in vitro adventitious bud techniques. Euphytica, 25: 11-19.

Buck, K.W. 1991. Virus resistant plants. In: Plant Genetic Engineering (ed. D. Grierson) pp 136-78

Buiatti, M. and Tesi, R. 1968. Gladiolus improvement through radiation induced somatic mutation. Publ. No. 1 Laboratoriodi Mutagenesie differentiamento del CNR, Pisa, pp. 1-6.

Buiatti, M., Molino, M. and Tesi, R. 1967. Prove di miglioramento genetico mediante radiationi ionitantio in una pianta da fiore a propagatione vegetative (gladiolo) Riv. Ortoflorofruttic Ital. 51 (6): 616-627.

Buiatti, M., Ragattini, R. and Togoni, F. 1965. Effects of gamma irradiation o gladiolus. Radit. Bot., 5: 97 98.

Burchi, G., Mercuri, A., Benedetti, L. De. and Giovannini, A. 1996. Transformation methods applicable to ornamental plants. Plant Tissue Culture Biotechnol., 2: 94-104.

Bush, A.l. and S.G., Pueppre 1991. Cultivar strain specificity between *Chrysanthemum morifolium* and *Agrobacterium tumifaciens*. Physiological and Molecular Plant Pathology, 39:309-323.

Bush, S.R., Earle, E.D. and Langhans, R.W. 1976. Plantlets from Petal segments, petal epidermis and shoot tips of the periclinal chimera, *Chrysanthemum morifolium* 'Indianapolis'. Am. J. Bot., 63: 729-737.

Cassells, A.C., Walsh, C. and Periappuram, C. 1993. Diplontic selection as a positive factor in determining the fitmness of mutants of Dianthus "Mystere' derived from x-irradiation of nodes in *in vitro* cuture. Euphytica, 70: 167-174.

Chakrabarty, D., Mandal, A.K.A. and Datta, S.K. 1999 Management of chimera through direct shoot regeneration from florets of chrysanthemum (*Chrysanthemum morifolium* Ramat). J. Hort. Sci. Biotech. 74(3): 293-296.

Chakrabarty, D., Mandal, A.K.A. and Datta, S.K. 2000. Retrieval of new coloured chrysanthemum through organogeneis from sectorial chimeras. Current Science 78: 1060-1061

Coen, E.S., Almeida, J., Robbins, T.P., Hudson, A. and Carpenter, R. 1988. Molecular analysis of genes determining spatial patterns in Anttirrhinum majus. In Temporal and spatial Regulation of plant genes (eds. DPS Verma and R.B Guilderberg), pp. 63-82.

Cohen, A. and Meridith, C.P. 1992. Agrobacterium-mediated transformation of lilium. Acta Hort., 325: 611-618.

Courtney-Gutterson, N., Firoozabady, E., Lemiex, C., Nicholas, J., Morgan, A., Robinson, K., Otten, A. and Akerboom, M. 1993. Production of genetically engineered colour modified Chrysanthemum plants carrying a homologous chaleone synthase gene and their field performance. Acta Hortic., 336:57-62.

Courtney-Gutterson, N., Napoli, C., Lemieux, C., Morgan, A., Firoozabady, E. and Robinson, K.E.P. 1994. Modification of flower colour in florist's Chrysanthemum: production of a white flowering variety through molecular genetics. Biotechnology, 12:268-271.

Craig, R. 1963. The inheritacne of several characters in the Geranium, *Pelargonium hortorum* Bailey. Thesis, Pennsylvania State Univ. Unviersity Park, pa, 71pp.

Custers, J.B.M., Van Eijk, J.P. and Spranaay, L.D. 1977. New developments in mutation breeding of vegetatively propagated ornamental crops with special reference to quantitative characters. In: Assoc. Euratom -ITAL (editor), The use of ionizing radiation in agriculture Proc. of workshop, Wagenigen, March 1976, pp. 505-510.

Das, P.K., Ghosh, P., Dube, S. and Dhuha, S.P. 1974. Induction of somatic mutations in some vegetatively propagated ornamentals by gamma irradiation. Technology (Coimbatore, India)1112(3): 185-186.

Das, P.K., Ghosh, P., Dube, S. and Dhuha, S.P. 1977. Improvement of some vegetatively propagated ornamentals by gamma irradiation. Indian J. Hortic., 34(2): 169-174.

Datta, S.K.1984.Cosmonaut–a new chrysanthemum cultivar evolved by gamma irradiation. J. Nucl. Agric. Biol., 13(40; 140.

Datta, S.K. 1985a. Colchi-mutation (C-mutation). Everyman's Science, XX(3): 70-72.

Datta, S.K. 1985b. Gamma ray induced mutant of a mutant chrysanthemum. J. Nuclear Agric. Biol. 14(4): 131-133.

Datta, S.K. 1986a. Effects of recurrent gamma irradiation on rose cv. 'Contempo'. J. Nuclear Agric. Biol. 15(2): 125-127.

Datta, S.K. 1986b. Thin layer chromatographic studies and evolution of somatic mutations in rose. The India Rose Annual, V: 41-48.

Datta, S.K. 1987. 'Colchi Bahar'–A new chrysanthemum cultivar evolved by colchi mutation. The chrysanthemum 43(1): 40.

Datta, S.K. 1988a. 'Agnishikha'–A new chrysanthemum cultivar evolved by gamma irradiation. Floriculture 9(11): 10.

Datta, S.K. 1988b. Chrysanthemum cultivars evolved by induced mutations at National Botanical Research Institute, Lucknow. The Chrysanthemum, 44(1): 72-75.

Datta, S.K. 1990a. Colchicine induced mutations in seed and vegetatively propagated plants. I. Indian Bot. Soc., 69: 261-266.

Datta, S.K. 1990b. Some basic informations for induction of somatic mutations in vegetatively propagated ornaemtnals. Advances in Horticulture and Foresty, 1, Article No. 33: 261-265.

Datta, S.K. 1990c. "Shabnam'–A gamma ray induced new chrysanthemum cultivar with new characters. The Chrysanthemum 46(4): 168-169.

Datta, S.K. 1991. Evaluation of recurrent irradiation on vegetatively propagated ornamental: Chrysanthemum. J. Nuclear Agric. Biol. 20(2): 81-86.

Datta, S.K. 1992a. Mutation studies on double bracted Bougainvillea at National Botanical Research Institute (NBRI), Lucknow, India. Mutation Breeding Newsletter, Issue No. 39, January: 8-9.

Datta, S.K. 1992b. Radiosensitivity of garden chrysanthemum J. Indian Bot. Soc. J. Indian. Bot. Soc., 71(I-IV): 283-284.

Datta, S.K. 1995a. Induced mutation for plant domestication: *Lantana depressa*. Proc. Indian. Sci. Acad B61 No. 1: 73-78.

Datta, S.K. 1995b. Role of Mutation Breeding in inducing desired genetic diversity in ornamentals for floriculture trade. Abstract published in Extended Synopsis, Abst. No. IAEA-SM-340/129 P, FAO/IAEA Internation Symposium on the 'Use of Induced mutation and molecular technique for Crop improvement', 19-23 June 1995, Vienna, Austria.

Datta, S.K. 1996. Effects of gamma irradiation on mutant genotypes: Chrysanthemum cultivar 'D-5' and its mutants. J. Indian Bot. Soc. 75(1-2): 133-134.

Datta, S.K. 1997a Exploring the chrysanthemum in India (editor S.K. Datta), EBIS, NBRI, Lucknow.

Datta, S.K. 1997b. Ornamental plants–Role of mutation. Daya Publishing House, Delhi, total pp. 219.

Datta, S.K. 1997c. Role of mutaion breeding for improvement of vegetatively propagated ornamentals. In: Plant Breeding Advances and *in vitro* culture (eds. Bahar A. Siddiqui and S. Khan), Dept. of Botany Aligarh Muslim University, Aligarh, CBS publishers and distributors, New Delhi: 144-158.

Datta, S.K. 1997d. Mutation studies on garden rose: A review. Proc. of Indian Natn. Sci. Acad., B63, No. 14, 2: 107-126.

Datta, S.K. 1998. Chrysanthemum germplasm at NBRI, Lucknow and search for novel genes. Applied Botany Abstracts, 18(1): 45-72

Datta, S.K. 2000. Mutation studies on garden chrysanthemum–A Review. Advances in Horticulture & Forestry (in press)

Datta, S.K. and Banerji, B.K. 1994. "Mahara Variegata'–A new mutant of bougainvillea. J. Nuclear Agric. Biol. 23(2): 114 -116.

Datta, S.K. and Banerji, B.K. 1997. Improvement of double bracted Bougainvillea through gamma ray induced mutations. Frontiers in Plant Science (Ed. Irfan A. Khan): 395-400.

Datta, S.K. and Gupta, M.N. 1981a. Cytomorphological, palynological and biochemical studies on control and gamma induced mutant of Chrysanthemum cultivar 'E-13'. SABRAO Journal, 134(2): 136-148.

Datta, S.K. and Gupta, M.N. 1981b. Cytomorphological, palynological and biochemical studies on control and gamma induced mutant of chrysanthemum cultivars 'D-5'. The Chrysanthemum, SABRAO J., 10(2): 149-161.

Datta, S.K. and Gupta, M.N. 1983a. Somatic flower colour mutation in Chrysanthemum cv. 'D-5' J. Nuclear Agric. Biol., 12(1): 22-23.

Datta, S.K. and Gupta, M.N. 1983b. Thin layer chrmatographic and spectrophotmetric analysis of flower colour mutations in roses. American Rose Annual: 102-108.

Datta, S.K. and Gupta, M.N. 1984. Effects of colchicine on rooted cutings of chrysanthemum. The chrysanthemum 40(5): 191-194.

Datta, S.K. and Gupta, M.N. 1985a. Treatment of budding eyes with colchicine for induction of somatic mutations in rose. The Indian Rose Annual, IV: 80-83.

Datta, S.K. and Gupta, M.N. 1985b. Mutation breeding on garden roses. American Rose Annual: 119-123.

Datta, S.K. and Gupta, M.N. 1987. Colchicine induced somatic flower colour mutation in chrysanthemum. Bangladesh J. Bot. 16(1): 107-108.

Datta, S.K., Banerji, B.K. and Gupta, M.N. 1985. 'Tulika'–A new chrysanthemum cultivar evolved by gamma irradiation. J. Nuclear Agric. Biol. 14(4): 160.

De Jong, J. and Custers, J.B.M. 1986. Induced changes in growth and flowering of chrysanthemum after irradiation and in vitro culture of Pedicels and Petal epdermis. Euphytica, 35: 137-148.

De Mol, W.E. 1949 twenty five years of tulip improvement by X rays. Pap. Micl. Acad. Sci. Arts. Lett., 35: 9-14.

de Vlaming, P., Corn, A., Fracy, E., Gerats, A.F.M., Maizonnier, D., Wiering, H. and Wijsman, H.J.M. 1984. Petunia Hybrida: A short description of the action of 91 genes, their origin and their map location. Plant Mol. Biol. Rept., 2: 21-42.

De, Mol, W.E. 1933. Mutation sowohl als modification durch Rontgenbestrahlung und die "Teilungshypothese". Cellule, 42: 149-160.

Deroles, S.C., Boase, M.R. and Konczak, I. 1997. Transformation protocols for ornamental plants. In R.L. Geneve *et al.* (eds). Biotechnology of Ornamental plants. CAB International pp, 87-119.

Desai, B.M. 1973. Report on mutation experiments on ornamentals at Bhabha Atomic Research Centre, Trombay, 2nd Workshop on Floriculture, Agric. Hort. Soc., Calcutta, Jan. 29-31st,

Desai, B.M. 1974. New cultivars of Perennial portulaca through gamma a radiation. Indian Hortic., 19(2); 19-23.

Dolgov, S.V., Mityshkina, T.U., Rukavtsova, E.B., Buryanov, Y.I, Vainstein, A. and Weiss, D. (1995). Production transgenic plants of *C.morifolium* Ramat with the gene of *B. thuringiensis* delta endotoxic. Acta Hortic., 420:46-47

Dooner, H.K. 1983. Co-ordinate genetic regulation of flavonoid biosynthetic enzymes in maize. Mol. Gen. Genet., 189: 136-141.

Dryagina, I.V. 1975. Gladioli forms obtained as a result of treatment with chemical mutagens. Vestn. Mosk. Univ., 6(2): 57-61.

Earle, E.D. and Langhans, R.W. 1974 b. Propagation of chrysanthemum in vitro 2. Production, growth and flowering of plantlets from tissue culture. J. Am. Soc. Hort. Sci., 99: 352-358.

Earle, E.D. and R.W. Langhans 1974 a. Propagation of chrysanthemum in vitro 1. Multiple plantlet from shoot tips and the establishment of tissue culture. J. Am. Soc. Hort. Sci., 99: 128-132.

Firoozabady, Noriega, E., C., Sondahl, M.R. and Robinson, K.E.P. 1991. genetic transformation of rose (*Rosa hybrida* cv. Royalty) via *Agrobacterium tumefaciens* in vitro, 27: 154 A.

Firoozabady, E., Moy, Y., Courtney, G. N. and Robinson, K. 1994. Regeneration of transgenic rose plants from embryogenic tissue. Biotechnology, 12(6):609-613.

Fukai, S., Jong, J.D., Rademaker, W., Nishio, T. and Dore, C. 1995. *Agrobacterium* medioated genetic transformation of Chrysanthemum. Acta Hortic., 392:147-152.

Gaul, H. 1958. Present aspects of induced mutations in plant breeding. Euphytica, 7:275-289.

Gaul, H. 1964. Mutations in Plant Breeding. Rad. Bot., 4: 155-232.

Gaul, H. 1965. Use of mutation for plant breeding in Europe. Proc. 6th Japan con. Radioisotopes, Tokyo, pp 843-860.

Gottschalk, W. and Wolff, G. 1983. Induced mutations in plant breeding. Springer-Vertlag, Berlin, Heidelberg, New York, Tokyo.

Grabowska, B. and Mynett, K. 1970. Induction of changes in garden tulips (*Tulipa hybr.* hort). under the influence of gamm rays ^{60}Co. Biul. Inst. Hodowli Aklim. Rosl.1-2: 89-92.

Graves, A.C.F. and Goldman, S.C. 1987. Agrobacterium tumefaciens-mediated transformation of the monocot genus gladiolus: detection of expression of T-DNA encoded genes: Journal of Biotechnolgy, 169: 1745-1746.

Griesbach, R.J. 1989. Selection of a dwarf Hemerocallis through tissue culture. Hort. Sci., 24: 1027-1028.

Griesbach, R.J. 2000. Gene gun speeds search for new orchid colours. [Downloaded from Internet. For personal contact, R.J. Griesbach, Floral and Nuresery Plants Research Unit, U.S. National Arboretum, Agric.Research Service, USDA, Beltsville, MD, (301) 504-6574].

Guo, J., Liu, Y., Tang, J. and Liu, C. 1983. Mutations in roses. J. nanjing Tech. Coll. Forest Prod., 2: 68-74.

Gupta, M.N. 1966. Induction of somtic mutation in soem ornamental plants. In Proc. All India Symp. Hortic., pp. 107-114.

Gupta, M.N. and Datta, S.K. 1983. Effect of colchicine on roses cv. 'Contempo'. The Indian Rose Annual III:51-54.

Gupta, M.N. and Datta, S.K. 1990. Role of induced mutation on rose breeding. The American Rose XXX: 17-18.

Gupta, M.N., and Shukla, R. 1974. Mutation breeding in Bougainvillea. Indian J. Genet., 34A: 1296-1299.

Gupta, M.N. and Nath, P. 1977. Mutation breeding in Bougainvillea II Further experiments with cvs. Partha and President Roosevelt. J. Nucl. Agric. Biol., 6(4): 122-124.

Gupta, M.N., Nath, P. and Datta, S.K. 1992. 'Mrinalini Stripe'–A new rose cv. evolved by gamma irradiation. The Indian Rose Annual X: 44-47.

Gupta, M.N., Swmiran, R and Shukla, R. 1974. Mutation breeding of tuberose Polianthes tuberosa) In: Symp.: Use of Radiations and Radioisotoped in studies of plant productivity, Pantnagar, pp. 169-179.

Gustafsson, A 1947. Mutation in agricultural plants.Hereditas, 33: 1-100.

Gustafsson, A. 1969. A study of induced mutations in plants.Induced mutations in plants. pp. 9-31, IAEA, Vienna.

Hacciys, B. and Hausner, G. 1974. Adventitvknospen und nicht-zygotische embryonene-Grundlagen and Anwendung. Vortr. 22, Vortragstg. Ges. Arzneipflanzen forsch Tubingen, pp. 1-10.

Hackett, W.P. and Anderson, J.M. 1967. Aseptic multiplication and maintenance of differentiated carnation shoot tissue derived from shoot apices. Proceedings of the American Society for Horticultural Science. 90: 365-369.

Halevy, A.H. and Mayak, S. 1981. Senescence post-harvest physiology of cut flowers. Part-2. Horticultural Review, 3: 59-143.

Heursel, J. 1980. Sportvorming bij Rhododendron simsii planch (Azalea indica L): een Synthase. L and bouwtijdschrift, 33: 985-994.

Hill, G.P. 1967. Morphogenesis of shoot primordia in culture stem tissue of a garden rose. Nature, 216: 596-597.

Hill, G.P. 1969. Shoot formation in tissue cultures of chrysanthemum 'Bronze Pride'. Physiol. Plant., 21: 386-389.

Hoekema, A., Huisman, M.J., Molendij, L., Van den Elzen, P.J.M. and Cornelissen, B.J.C. 1989. The genetic engineering of two commercial potato cultivars for resistance to potato virus X. Bio/Technology, 7: 273-278.

Holton, T.A., Kovacic, F., Tanaka, Y., Lester, D.R., Hyland, C., Gloster, S., Michael, M.Z., Perilli, T., O'Connor, E., Nakamura, N., Caesar, C., Tsuda, S., Menting, J.G.T., Stvenson, T.W., Comish, E. and Dalling, M. J. (1992). Isolation and expression of two cytochrome P-450 genes controlling flower colour in petunia. 14th Annual Conference on the Organization and Expression of the Genome. Lome, Victoria.

Hutchinson, J.F., Kaul, V., Mahaeswaran, G., Moran, J.R., Graham, M.W. and Richards, D. 1992. Genetic improvement of floricultural crops using biotechnology. Aust. J. Bot., 40: 765-787.

James, J. 1983. New roses by irradiation: an update. Am. Rose. Ann., 99-101.

Janssen, B.J. and Gardner, R.C. 1989. Localised transient expression of GUS in leaf disces following co-cultivation with Agrobacterium. Plant Mol. Biol., 14: 61-72.

Jefferson, R. 1987. Assaying chimeric genes in plants: The GUS gene fusion system: Plant molecular Biology Reporter, 5: 387-405.

Jenks, M.A. and Feldmann, K.A. 1997. T-DNA insertion mutagenesis for improvement of ornamentals. In: Biotechnology of ornamentals plants, pp. 185-197. (eds. R.L. Geneve, J.E. Preece and S.A. Merkle). Wallingford: C.A.B. Internationsl.

Jerzy, M. 1990. *In vitro* induction of mutations in chrysanthemum using x–and y-radiation. Mutat. Breed. Newsl., 35: 10.

Jerzy, M. and Lunomski, M. 1991. Adventitious shoot formation on ex in vitro derived leaf explants of *Gerbera jamesonii*. Scientia Hort., 47: 115-24.

Jerzy, M. and Zalewska, M. 1996. Polish cultivars of *Dendranthema grandiflora* Tzvelev and *Gerbera jamsonii* Bolus bred in vitro by induced mutation. Mutant Breed Newsl., 42: 19.

Jong, J.D., Rademaker, W. and Wordragen, M.F. 1993. Restoring adventitious shoot formation on Chrysanthemum leaf explants following co-cultivation with *Agrobacterium tumifaciens*. Plant cell Tissue and organ Cult., 32:263-270.

Jong, J.D., Mortens, M.M.J., Radermaker, W. and De Jong, J. 1994. Stable expression of the GUS reporter gene in Chrysanthemum depends on binary plasmid T-DNA. Plant Cell Report, 14:59-64.

Kaicker, U.S. 1985. The creation of new roses at the Indian Agricultural Research Institute–A Silver Jubilee. The Indian Rose Annual, IV: 50-58.

Kaul, V., Miller, M.R., Hutchinson, J. F. and Richards, D. 1990. Shoot regeneration from stem and leaf explants of *Dendranthema grandiflora* Tzvelev (Syn. *Chrysanthemum morifolium* Ramat). Plant Cell Tissue Organ Cult., 21: 21-30.

Kawai, T. and Amano, E. 1991. Mutation breeding in Japan. In: Plant Mutation Breeding for Crop Improvement. pp. 47-66.

Kende, H. 1989. Enzymes of ethylene biosyntheis. Plant Physiology, 91: 1-4.

Khalid, N., Davey, M.R. and Power, J.B. 1989. An assessment of somaclonal variation in *Chrysanthemum morifolium*: The generation of plants of potential commercial value. Scientia Hort., 38, 287-294.

Kim Mi Jung, Kim, Y.J., Kim, M.U and Kim, Y.H. 1998. Plant regeneration and flavonoid 3′5′ hydroxylase gene transformation of *D. zawadskii* and *D. indicum*. J. Korean Soc. Hort. Sci., 39:355-359.

Kim, J.Y., Park, S.J., Um, B.Y., Pak, C.H., Chung, Y.S. and Shin, J.S. 1998. Transformation of Chrysanthemum by *A.tumefaciens* with three different types of vectors. J. Korean Soc. Hort. Sci., 39: 360-366.

Kintzios, . S., Manos, C. and Makri, O. 1999. Somatic embryogenesis from mature leaves of rose. Plant cell Reports, 18(6): 467-472.

Klimenko, Z.K. and Zykov, K. I. 1977. Effect of exposures of garden rose cuttings to gamma radiation on development and morphological variability. Radiobiology, 17(6): 152-155.

Koes, R.E., Spelt, C.E., Reif, H.J., vander Elzen, P.J.M., Veltkamp, E. and Mol, J.N.M. 1986. Floral tissue of *Petunia hybrida* (V30) expresses only one member of chalcone synthase multigene family. Nucleic Acid Research, 14: 5229-39.

Konzak. C.F. 1956. Induction of mutation for disease resistance in cereals. Brookhaven Symp. Biol., 9:157-176.

Konzak. C.F. 1957. Genetic effects of radiation on higher plants. Q.Rev. Biol., 32: 27-45.

Kurle, R., Meldrays, J., Druka, A. and Linde, S. 1993. Transient expression of a foreign gene in Chrysanthemum protoplasts. Latvijas Zinatnu Akademijas Vistis -B-dala dabaszinatnes.1:62-64.

Laneri, U., Franconi, R. and Altavista, P. 1990. Somatic mutagenesis of *Gerbera jamesonii* Hybr. irradiation and in vitro culture. Act Hort., 280: 395-402.

Langveld, S.A., Derks, A.F.L.M., Gerrits, M.M., Boonekamy, P.N. and Bol, J.F. 1994. Tranformation of lily by Agrobacterium VIIIth International Congress of Plant Tissue and Cell Culture, Abst. M-23.

Larkin, P.J. and Scowcroft, W.R. 1981. Somaclonal variation–a novel source of variability from cell culture for plant improvement. Theoretical and applied Genetics, 60: 197-214.

Lawton, K.A., Huang, B., Goldsbrough, P.B. and Woodsen, W.R. 1989. Molecular cloning and characterization of senescence-related genes from flower petals. Plant Physiol., 90: 690-696.

Ledger, S.E., Deroles, S.C. and Given, N.K. 1991. Regeneration and *Agrobacterium* mediated transformation of Chrysanthemum. Plant Cell Reports, 10:195-199.

Leitner-Dagan, Y., A. Izhaki, A., Ben-Nissan, G., Borochov, A. and Weiss, D. 1999. GA-induced gene expression in petunia flowers. In: Plant Biotechnology and In Vitro Biology in the 21st century: 162-172. Kluwer Academic Publishers.

Lemieux, C.S., Firoozabady, E. and Robinson, K.E.P. 1990. Agrobacterium–mediated transformation of chrysanthemum. proceedings of the EUCARPIA Symposium on Integration of in vitro techniques in ornamental plant breeding, pp 150-155.

Lowe, J.M., Davey, M.R., Pown, J.B. and Blundy, K.S. 1993. A study of some factors affecting Agrobacterium transformation and plant regeneration of *Dendranthema grandiflora* Tzvelev (Sm. *Chrysanthemum morifolium* Ramat).. Plant Cell Tissue & Organ Culture, 33: 171-180.

Lu, C.Y., Nugent, G. and Wardley, T. 1990. Efficient, direct plant regeneration from stem segments of chrysanthemum. (*C. morifolium* Ramat. cv. Royal purple). Plant cell reports, 8: 733-736.

Lu, C.Y., Nugent, G., Wardley-Richarddson, T., Chandla, S.F., Young, R., Dalling H.J. 1991 Agrobacterium mediated transformation of carnation (Dianthus caryophyllus L). Bio/Technology, 9: 864-68.

Mahna, S.K. and Garg, R. 1989. Induced mutation in *Petunia nyctaginiflora* Juss. Biol. Plant., 31: 152-155.

Malaure, R.S., Barday, G., Power, J.B. and Davey, M.R. 1991a. The production of Novel plants from florets of *Chrysanthemum morifolium* using tissue culture 1. Shoot regeneration from ray florets and somaclonal variation exhibition by the regenerated plants. J. Plant. Physiol., 139: 8-13.

Malaure, R.S., Barday, G., Power, J.B. and Davey, J.B.1991b. The production of Novel plants from florets of *Chrysanthemum morifolium* using tissue culture 2. Securing natural mutations (sports) J. Plant. Physiol., 139: 14-18

Malauszynski, M., Van Zanten, L., Ashri, A., Brunner, H., Ahloowalia, B., Zapata, F.J., and Weck, E. 1995. Mutation techniques in plant breeding. In: Induced Mutation and Molecular Techniques for Crop Improvement. Proc. of a Symp., Vienna 19-23 June, IAEA, Vienna.

Mandal, A.K.A., Chakrabarty, D. and Datta, S.K. 2000a. In vitro development of novel flower colour through management of induced mutation. Euphytica, 114: 9-12

Mandal, A.K.A., Chakrabarty, D. and Datta, S.K. 2000b. Use of in vitro techniques in mutation breeding of Chrysanthemum. Plant Cell Tissue & Organ Culture 60(1): 33-38.

Marchant, R., Davey, M.R, Lucas, J.A. and Power, J.B. 1996. Somatic embryogenesis and plant regeneration in floribunda rose cv Trumpeter and Glad Tidings. Plant Sc., 120: 95-105.

Marchant, R., Davey, M.R., Lucas, J.A., Lamb, C.J., Dixon, R.A. and Power, J.B. 1998 Genetic engineering of rose for resistance to blackspot. In Tree Biotechnology towards the Millennium. pp 291-299

Marchant, R., Power, J.B., Davey, M.R. and Lucas, J.A. 1998 Biolistic transformation of rose (*R. Hybrida* L). Annals of Botany, 81: 109-114.

Martin, Carpenter, C.R., Loen E.S., and Gerats, A.T.M. 1987. The content of floral pigmentation in *Antirrhinum majus*, pp. 19-52. In: H. Thomas and D. Grierson (eds). Developmental mutants of higher plants. Bio. Technology, Cambridge University Press, Cambridge, U.K.

Matsubara, H., Iba, S., Oka, M. and Meshitsuka, G. 1965. Effects of gamma-irradiation on tulip II. Effects on various stages of development. Tokyo Metrop. Isot. Centre. Annu. Rep., 21 (1963): 157-162.

Matsuda, Y., Toyoda, H., Nakahara, T., Ogata, Y. and Ouchi, S. 1996. Basic study for application of microprojectile bombardment to the production of transgenic rose plants, Bulletin of the Institute for Comprehensive Agricultural Sciences, Kinki University, 4: 23-28.

May, R.A. and Trigiano, R.N. 1991. Somatic embryogenesis and plant regeneration from leaves of *Dendranthema grandiflora*. J. Amer. Soc. Hort. Sci., 116(2): 366-371.

Messeguer, J. and Mele, E. 1994. Factors influencing genetic transformation in carnation. VIIIth International Congress of Plant Tissue & Cell Culture, Abst. S7-77.

Meyer, P. Heidmann, I., Forkmann, G. and Seeler, H. 1987. A new Petunia flower colour variety generated by transformation of a mutant with a maize gene. Nature, 330: 677-78.

Meynet, J and Sibi, M. 1984. Haploid plants from in vitro culture of unfertilized ovules in Gerbera jamesonii. Zeitschriftfur Plan Zanzuchtung, 93: 78-85.

Michael, M.Z., Savin, K.W., Baudinette, S.C., Graham, M.W., Chandler, S.F., Lu, C.Y. Caesar, C., Gautrais, R., Young, R., Nugent, G.D., Stevenson, K.R., O'Corner, E.L.J., Cobbett, C.S. and Wrnish, E.C. 1993. Cloning of ethylene biosynthetic genes involved in petal senescence of carnation and petunia and their antiscence expression in transgenic plants. In: Pech. J.C. *et al.* (eds).. Cellular & Molecular Aspects of the plant harmone Ethylene. Kluwer Acad. pp. 298-303.

Miller, R.M., Kaul, V., Hutchinson, J.F. and Richards, D. 1991. Adventitious shoot regeneration in carnation *Dianthus caryophylus*) from axillary bud explant. Annals of Botany, 67: 35-42.

Miller, R.M., Kaul, V., Hutchinson, J.F., Mahesarn, J.F. and Richards, D. 1991. Shoot regeneration from fragmented flower buds of carnation in carnation (*Dianthus caryophylus*). Annals of Botany, 67: 35-42.

Mol, J.N.M., Stuitje, A.R., Van der Krol, A. 1989. Genetic manipulation of floral pigmentation genes. Plant Molecular Biology, 13: 287-294.

Mor, Y, Spiegelstein, H and Halvey, A.H. 1983 Inhibition of ethylene biosyntheis in Carnation petals by cytokinni. Plant Physiology, 71: 541-46.

Murashige, T., Sepra, M. and Jones, J.B. 1974. Clonal multiplication of Gernbera through tissue culture. Hort. Science, 9: 175-180.

Nagatomi, S., Degi, K., Yamaguchi, M., Miyahira, E., Sakamoto, M. and Takaesu, K. 1993. Six mutant varieties of different flower colour induced by floral organ culture of chronically irradiated chrysanthemum plants. Technical News-Institute of Radiation Breeding No. 43: 1-25.

Nagatomi, S., Kamiyo, M., Narusawa, H., Iwasaki, T., Okazaki, T. and Maruta, Y. 1996. Three mutant varieties in Eustoma grandiflorum through in vitro culture of chronic irradiated plants. Technical News–Institute of Radiation Breeding No. 53: 1-2.

Nakajima, N., Mor, H., Yamazaki, K and Imaseki, H 1990. Molecular cloning and sequence of a complementary DNA coding ACC synthase induced by tissue wounding Plant and cell physiology, 31: 1021-29.

Naylor, *I.E.* and Johnson, B. 1973. A histological study of vegetative reproduction in *Saintpaulia ionatha*. Am. J. Bot., 24: 673-678.

Nezu, M. 1967 Tulip breeding by bud sports induced by gamma rays. Toyama Agric. Exp. Stn. Spec. Rep., No. 7, pp. 1-74.

Nikaido, T. and Onozawa, Y. 1989. Establishment of a non-chimeric flower colour mutation through in vitro culture of florets from a sport in chrysanthemum with special reference to the genetic background of the mutation line obtained. Scientific Reports of the Faculty of Agriculture, Ibaraki University, 37: 63-69.

Noriega, C. and Sondhal, M.R. 1991. Somatic embryogenesis in hybrid tea roses. Bio/technology, 9: 991-993.

Nugent, G., Wardley-Richardson, T. and Lu, C.Y. 1991. Plant regeneration from stem and petal of carnation (*Dianthus caryophyllus* L).. Plant Cell Reports, 10: 477-480.

Nybom, N. and Koch, A. 1965. Induced mutations and breeding methods in vegetatively propagated plants, The use of induced mutations in plant breeding (Rep. FAO/IAEA Tech. Meeting, Rome 1964, Pergamon Press, Oxford: 661-678. Rad. Breed. 2:1-102.

Ohba, K. 1971. Studies on the radiation breeding of forest trees. Bul. Inst. Radiat. breed. 2, Ohmiya, Ibaraki: 1-102.

Orlikowska, T., Nowak, E., Marasek, A, Kucharska, D. 1999. Effects of growth regulators and incubation period on in vitro regeneration of adventitious shoots from gerbera petioles. Plant Cell, Tissue and Organ Culture, 59: 95-102.

Ovadis, M., Zuker, A., Tzfira, T, Ahroni, A., Shklarman, E., Scovel, G., Itzhaki, H., Ben-Meir, H. and Vain Stein, E. 1999. Generation of transgenic carnation plants with novel characteristics by combining microprojectile bombardment with Agrobacterium tumefaciens transformation. In: Plant Biotechnology and In Vitro Biology in the 21st Century: 189-192. (eds. A. Altman *et al*).

Pavingerova, D., J. Dostal, R. Biskova and V. Benetka 1994. Somatic embryogenesis and *Agrobacterium* mediated transformation of Chrysanthemum. Plant Science, 97:95-101.

Pellegrineschi, A., Damon, J.P., Voltorta, N., Paillard, N. and Tepfer, D. 1994. Improvement of ornamental characters and fragrance production in lemon scented geranium through genetic transformation by Agrobacterium rhizogenus. Bio. Technology, 12: 64-68.

Pierik, R.L.M. 1991. Commercial micropropagation in western europe and Israel. In micropropagation: Technology and Application (eds. P. Deberg and R.H. Zinmerman) pp: 155-165.

Pierik, R.L.M., Steegmans, H.H.M. Verhaegh, J.A.M. and Wouters, A.N. 1982. Effect of cytokinins and cultivar on shoot formation of Gerbera jamesonii in vitro. Neth. J. Agric.. Sci., 30: 341-346.

Pierik, J.L.M., Jansen, A., Maasdam, and Binnendijk, C.M.. 1975. Optimization of Gerbera plnatlet production from excised capitulum explants. Scientia Horticulturae, 3: 351-357.

Powell-Abel, P., Nelson, R.S., De, B., Hoffmann, N., Rogers, S.G., Fraley, R.T. and Beachy, R.N. 1986. Delay of disease development in transgenic plants that express the tobacco mosaic virus coat protein gene. Science, 232: 738-743.

Privalov, G.F. 1967. Experimental mutation in woody plants, Induced mutations and their utilization. Proc. Symp. Erwin-Bauer-Gedachtnisvorlesungen IV., Gatersleben, 1966, Akademie-Verlag, Berlin: 383-386.

Raev, R.T. 1984. Mutation breeding of Kazanlik essential oil-bearing rose (R. damascena Hill) Rastenievud. Nauki, 21(8); 92-99.

Raguvanshi, S.S. and Singh, A.K. 1979. Gamma ray induced mutations in diploid and autotetraploid perinnial *Portulaca grandiflora* Hook. Indian J. Hortic., 36: 84-87.

Raguvanshi, S.S. and Singh, A.K. 1980. Autoploid radiosensitivity with special reference to *Portulaca grandiflora* L. In: S.S. Bir (editor), Recent researches in Plant Sciences. Kalyani Publishers, New Delhi, pp. 347-352.

Renou, J.P., Brochard, P. and Jalonz, R. 1993. Recovery of transgenic Chrysanthemum after hygromycin resistance selection. Plant Science. 89:185-197.

Reynoird, J.P., Chriquai, D., Noin, M., Brown, S., Marie, D.1993. Plant regeneration from in vitro leaf culture of several Gerbera species. Plant Cell, Tissue and Organ Culture, 33: 203-210.

Richter, A. and Singleton, W.R. 1955. The effect of chronic gamma radiation on the production of somatic mutations in carnations. Proc. Natl. Acad. Sci. USA, 41(5): 295-300.

Robinson, K.L.P. and Firoozabady, E. 1993. Transformation of floricultural crops. Scientia Hort. 55: 83-99.

Rout, G.R., Debata, B.K. and Das, P. 1992. In vitro regeneration of shoots from callus cultures of *Rosa hybrida* L cv. Landora. Indian Journal of Experimental Biology, 30: 15-18.

Ruffonie, B.J. and Bmassabo, F. 1991. Tissue culture of Gerbera jamesonii hybrida. Acta Hotic., 289: 147-148.

Schum, A. and Preil, W. 1998. Induced mutations in ornamental plants. In S.M. Jain, D.S. Brar, B.S. Ahloowalia (eds)., Somaclonal variation and induced mutations in crop improvement, pp. 333-366.

Seiichi, F., Jong, J. D. and Rademaker, W. 1995. Efficient genetic transformation of chrysanthemum (*D. grandiflorum* Ramat. cv. Kitamura) using stem segments. Breeding Sci., 45(2):179-184

Sherman, J.M., Moyer, J.W. and Daub, M.E. 1998. A regeneration and *Agrobacterium* mediated transformation system for genetically diverse chrysanthemum cultivars. J. Amer. Soc. Hort. Sci., 123:189-194.

Sherman, J.M., Moyer, J.W., Daub, M.E. and Kuo, C.G. 1996. Genetically engineered resistance to tomato spotted wilt virus in Chrysanthemum–a model system for protection in ornamental crops. Acta Hortic., 431:432-441.

Short, K.C. and Roberts, A.V. 1991. Rosa sp. (rosea): In vitro culture, micropropagation and the production of secondary products. In Bajaj Y.P.S. (ed) Biotechnology in Agriculture & Forestry, Vol. 15, Medicinal and Aromatic Plants III, Springer-Verlag, Berlin, pp. 376-397.

Shukla, R. and Datta, S.K. 1993. Mutation studies on early and late varieties of garden chrysanthemum. J. Nuclear Agric. biol., 23(3): 146-147.

Sigurbjornsson, B. and Micke, A. 1969. Progress in mutation breeding. Induced mutations in plants (Proc. symp. pullman, 1969), IAEA, Vienna: 673-698.

Silvy, A. and Mitteau, Y. 1986. Diversification des varietes doellet par traitement mutagene. In: Nuclear Techniques and in vitro culture for plant improvement, pp. 395-405, Vienna: IAEA.

Simard, M.H., Michaux-Ferriere, N. and Silvy, A. 1992. Variants of carnation (*Dianthus caryophylus* L). obtained by organogenesis from irradiated petals. Plant Cell, Tissue and Organ culture, 29: 37-42.

Sitbon, M. 1981. Production of haploid Gerbera jamesonii plants by in vitro culture of unfertilized ovules. Agronomic, 1: 807-12.

Skirvin, R.M. and Janick, J. 1974. Calliclones in geranium. Hort.Science, 9(3): 270.

Skirvin, R.M. and Janick, J. 1976 Tissue culture-induced variation in scented Pelargonium spp. Journal of the American Society for Horticultural Science, 101 (3): 281-290.

Smith, H.H. 1958. Radiation in the production of useful mutations. Bot. Rev., 24:1-24.

Sommer, H., Beltran, J.P., Huiser, P., Pape, H., Lonnig, W.E., Saedler, H. and Schwarz-Sommerz 1990. Deficiens, a homeotic gene involved in the control of flower morphogenesis in Antirrihinuss majus, the protein shows homology to transcription factors. EMBO Journal, 9:605-613.

Sparrow, A.H., Sparrow, R.C. and Schairer, L.A. 1980. The use of X rays to induce somatic mutations in Saintpaulia. Afr. Violet Mag., 13: 32-37.

Stimart, D.P., Ascher, P.D. and Zagorski, J.S. 1980. Plants from callus of the interspecific hybrid Lilium 'Black Beauty'. HortScience, 15: 313-315.

Stube, H. (1959). Some results and problems of theoretical and applied mutation research. Indian J. Gent. Plant Breed., 19:13-29.

Sutter, E. and Langhans, R.W. 1981. Abnormalities in chrysanthemum regenerants from long term cultures, Ann. Bot., 48, 559-568.

Swaminathan, M.S. 1961. Effects of diplontic selection on the frequency and spectrum of mutations induced in polyploids following seed irradiation. In: The effect of ionizing radiation on seeds and their significance for crop improvements. Kaernferring, Vienna 1, Austria, Intnl. At. Energy Agency: 279-288.

Swaminathan, M.S. 1963. Evaluation of the use of induced micro and macro mutations in breeding of polyploid crop plants. Proc. Symp. Energy nuclear agric. Rome, pp 243-277.

Swaminathan, M.S.1971. Mutation breeding and agricultural progress. J.Indian Bot. Soc., 50A: 416-429.

Takatsu, Y., Tomotsune, H., Kasumi, M. and Sakuma, F.1998. Difference in adventitious shoot regeneration capacity among Japanese Chrysanthemum (*Dendrathema grandiflora* Ramat. kitamura cultivars) and the improved protocol for *Agrobacterium* mediated genetic transformation. J. Japan. Soc. Hort. Sci., 67: 958-964.

Tasuda, S., Suzuki, K., Xue, H., Tanaka, Y. Fukui, Y., Mizutani, M.F., Katsumoto, Y. and Kusumi, T. 1999. Molecular breeding of flower colour of *Torenia hybrida*. In: A. Altman *et al.*(eds)., Plant Biotechnology and In Vitro Biology in the 21st Century, pp. 613-616.

Tumer, N.E., K.M.. O'Connell, R.S. Nelson, P.R. Sanders, R.N., Beach R.T. and D.M. Shah 1987. Expression of alfalfa mosaic virus caot protein confers cross-protection in transgeneic tobacco and tomato plants. EMBO J. 6: 1181-1188.

Urban, L.A., Sherman, J.M., Moyer, J.W. and Daub, M.E. 1994. High frequency shoot regeneration and *Agrobacterium* mediated transformation of Chrysanthemum. Plant Science., 98:69-79.

Vain Stein, A., Fisher, M. and Ziv, M. 1993. Applicability of reporter gene to Carnation transformation. Hort. Sci. 28: 1122-1124.

Van der Krol, A.R., Lenting, P.E., Veenstra, J., van der Meer, J.M., Koes, R.E., Gerats, A.G.M., Mol, J.N.M. and Stuitje, A.R. 1988. An antisense chalcone synthase gene in transgenic plants inhibits flower pigmentation. Nature, 333: 866-869.

Van der Straeten, D., Van Wiemeersch, L., Goodman, H.M. and Van Montagu, M. 1990. Cloning and sequence of two different cDNAs encoding 1-aminocyclopropane-1-carboxylate synthase in tomato. Proceedings of the National Academy of Sciences, U.S.A., 87: 4859-4863.

Van Eijk, J.P. and Eikelnboom, W. 1981a. Selectie bij veredeling van tulpen. Hobaho, 54(50): 6-8.

Van Eijk, J.P. and Eikelnboom, W. 1981b. Selectie bij de veredeling van de tulp (II) Bloembollencultuur, 91(39): 1054-55.

Van Groenestijn, J.E. and J., Van Tuyl 1983. Straks green leliebelichting meer nodig. Hoopvolle resultaten van veredeling Op lichtbchoefle. Vakbl Bolemisterij, 38: 30-31.

Van Tunen, A.J. and Mol, J.N.M. 1991 Control of flavonoid synthesis and manipulation of flower colour. In development regulators of plant gene expression (eds. D. Grierson), pp. 94-130.

Venkatchalam, P. and Jayabalan, N. 1994a. New flower colour mutants evolved through gamma irradiation in *Zinnia elegans* Jacq. J. Mendel. 11: 35.

Venkatchalam, P. and Jayabalan, N. 1994b. Effect of gamma irradiation on flower structure of *Zinnia elegans* Jacq. J. Swamy Bot. Cl., 11: 58.

Walther, F. and Sacer, A. 1986. In vitro mutagenesis in Gerbera jamesonii. In: Genetic Manipulation in plant breeding, pp. 555-562 (eds. W. Horn, C.J. Jensen, W. Odenbach and O. Schieder), Berlin: Walter de Gruyter.

Walther, F. and Sauer, A. 1991. Split dose irradiation of *in vitro* derived microshoots. An effective procedure for increasing mutability. In: Plant Mutation Breeding for Crop Improvement, : 343-353, Vienna. IAEA.

Wang, L.Q. 1991. Induced mutation for crop improvemnt in China. In: Plant Mutation Breeding for crop improvement, pp. 9-32, Vienna IAEA.

Wang, H. and Woodson, W.R. 1991. A flower senescence related mRNA from carnation shares sequence similarity with fruit ripening-related m-RNAs involved in ethylene biosyntheis. Plant Physiology, 96: 1000-01.

Wang, K., Herrera, E.A. and Van Montagci, M. 1990. Over expression of vir D1 and Vir D2 gene in *Agrobacterium tumefaciens* enhances T-complex formation and plant transformation. J. of Bacteriology, 172: 4432-4440.

Wiering, H. and de Vlaming, P. 1984. Inheritance and biochemistry of pigments. In Petunia. (ed. K.C. Sink), pp. 49-67.

Wilmink, A., Van den Var, B.C.E. and Dons, J.J.M. 1992. Expression of GUS gene in the monocot tulip after introduction by particle bombardment and Agrobacterium. Plant Cell Reports, 11: 76-80.

Wilmink, A., Van den Var, B.C.E. and Dons, J.J.M. 1994. A transformation procedure for tulip. IVth International Congress of Plant molecular biology 19-24 June, Amsterdam, The Netherlands, Abst. 2038.

Winefield, C., Davies, K., Schwinn, K., Deroles, S., Boulton, G., Miller, R, Boase, M. and Konczac, I.I. 1994. Genetic engineering for pelargonidin production in chrysanthemum. IVth Annual Queenston Molecular Biology Meeting, Queenstown, New Zealand, Abs. 072.

Woodson, W.R. 1991. Biotechnology of floriculture crops. Hort. Science: 26: 1029-1033.

Woodson, W.R. and Goldsbrough, P.B. 1989. Genetic transformation of carnation using *Agrobacterium tumefaciens*. Hort. Science, 24: 80.

Woodson, W.R., Lawton, K.A., Meyer, R.C., Ragnothama, K.G. and Goldbrough, P.B. 1990. Regulation of gene expression in sensing carnation petals. In Horticultural Biotechnology. (ed. A.B. Bennett and S.D. O'Neill), pp. 203-12.

Wordragen, M.F. and G. Howe 1990. The use of endotoxin genes from *B.thuringiensis* to introduce insect resistance in Chrysanthemum callus. Proceedings of Integration of in vitro techniques in ornamental plant breeding. 145-149.

Wordragen, M.F., Jong, J.D., Schornagel, M.J. and Pons, H.J.M. 1992. Rapid screening for host bacterium interactions in *Agrobacterium* mediated gene transfer to Chrysanthemum.Plant Science, 81:207-214.

Yanofsky, M.F. Ma, H., Bowman, J.L., Drews G.N., Feldmann, K.A. and Meyerowitz, E.M. 1990. The proteins encoded by the Arabidopsis homeotic gene AGAMOUS resembles transcription factors. Nature, 346: 35-39.

Yepes, L.M., Mittak, V., Slightom, J.L., Pang, S.T., Gonsalves, D., Fischer, G. and Angarita, A. 1999. *Agrobacterium tumefaciens* versus biolistic-mediated transformation of the Chrysanthemum cv Polaris and Golden polaris with nucleocaspid protein genes of three tospovirus species. Acta Hortic., 482:209-218.

Younis, S.E.A. and Borham, J.M. 1975. The effects of gamma irradiation on *Polianthes tuberosa.* Egypt J. Bot., 18(1&3): 205-

Chapter 14

Combination of Classical and Modern Methods for the Development of New Ornamental Varieties

A.K.A. Mandal & S.K. Datta

Floriculture Section, National Botanical Research Institute, Lucknow– 226001 (U.P.), India

ABSTRACT

In mutation breeding of vegetatively propagated ornamentals, formation of chimera is the main problem. A small sector of a mutated branch or flower cannot be isolated using the conventional propagation techniques that resulted in loss of new flower colour/shape mutants. In our laboratory an efficient system for isolation of such flower colour/shape chimera has been developed in chrysanthemum. In this system, mutated ray florets has been cultured on Murashige and Skoog (MS) medium containing growth regulators. Direct shoots has been regenerated from these ray florets. These shoots after rooting were transferred to the field where they have flowered true-to-explant floret colour/shape. Following this system we have isolated several mutants in pure forms.

INTRODUCTION

Floriculture industry must respond to frequently changing consumer demand. Changes in aesthetic value create an everincreasing demand for flowers which have unusual or novel characteristics. An ornamental variety having much demand today may be abandoned by consumers tomorrow. Therefore, development of newer and newer cultivars and their fast marketing are major challenges in floriculture industry. Manipulating the existing gene pool by conventional breeding and selection, induced mutation breeding and selection of natural mutations (sports) have been the main techniques of creating new cultivars. Use of these techniques in a planned way have developed an abundant array of attractive cultivars in different ornamental plants. These techniques, particularly mutation breeding, will undoubtedly continue to be of immense value. But the main drawback of mutation breeding is the formation of chimera. The formation of chimera after mutagen treatment is common and remains main constraint in mutation breeding of vegetatively propagated crops (D'Amato, 1965). In vegetatively propagated plants, mutation appears as a chimera after treatment with physical and/or chemical mutagens. In chimeric tissues, mutated cells are present along with the normal cells. The size of the mutant sector varies from a narrow streak on a petal to the entire branch. When a portion of a branch or an entire branch is mutated, the mutant tissue can be isolated, on the other hand, a small sector of a mutated branch or flower cannot be isolated using the available propagation techniques. Therefore, many new flower colour/shape mutants are lost due to lack of regeneration system from small mutated sectors either *in vivo* or *in vitro*. Hence it is very important to develop tissue culture technique to retrieve mutants from chimeric tissues.

Chrysanthemum (*Chrysanthemum morifolium* Ramat) is one of the most important cut flowers in floriculture trade. In chrysanthemum, a large number of chimeric flower colour/shape mutants are lost every year due to non-availability of regeneration system from small sectors. In our laboratory a very efficient protocol for direct shoot organogenesis from ray florets of several chrysanthemum cultivars has been developed. Using this regeneration system we have isolated several mutants in pure form from chimeric form (Chakrabarty *et al.*, 1999; Mandal *et al.*, 2000;

Datta *et al*., 2001). This procedure of chimera management has been presented here.

Materials and Methods

Rooted cuttings of chrysanthemum ((*Chrysanthemum morifolium* Ramat). cvs. Purnima, Colchi Bahar, Puja and Maghi are treated with different doses of gamma rays (Cobalt 60 radiation source at dose rate of 19 sec/krad) during the normal planting season. Twenty cuttings were irradiated at each level. The treated and control cuttings were planted in earthen pots and allowed to flower. In case of the cv. Kasturba Gandhi, flower colour mutation appeared as natural sport in the germplasm collection of NBRI, Lucknow. The mutation appeared in only one flower where few florets developed yellow colour.

Flower colour chimera appeared as sectorial chimera. Therefore, the mutated ray florets were collected from flower heads and washed in running tap water for 15 min, followed by 5 min wash in liquid detergent and then in single distilled water (4-5 times). The florets were then disinfected in 70 per cent ethanol for 30 sec followed by surface sterilization with $HgCl_2$ (0.1 per cent , w/v) for 2 min and then thoroughly washed in sterile distilled water (4-5 times). The whole florets were used as explants. The explants were cultured on Murashige and Skoog,s medium (Murashige and Skoog, 1962) containing 0.8 per cent agar, 3 per cent sucrose and plant growth regulators (1-naphthaleneacetic acid–NAA, kinetin and 6-benzyladenin–BA). Medium pH was adjusted to 5.6 before autoclaving at 121°C for 15 min. all the cultures were incubated under a 16 h photoperiod (36 µmol/m^2/s) at 25 1°C. Shoots of 2-3 cm in length were rooted on half strength MS medium supplemented with 1.5 per cent sucrose, 0.8 per cent agar and 0.02 mg/l NAA. Rooted shoots were transferred to plastic pots containing mixture of sand and soil (1:1) and kept under high humidity for one week. After another two weeks of hardening plantlets were transferred to the field and grown up to flowering. The isolated mutants were maintained by vegetative propagation (cuttings) method during successive years.

Results and Discussion

Flower colour/ray floret shape mutation was observed in treated population of all the experimental cultivars (Table 14.1). No mutation in flower was observed at 1 krad treatment where as in 1.5, 2.0 and 2.5 krad treatment flower colour / floret shape mutation was observed in different frequencies in different cultivars. Flower mutation in all the cultivars were chimeric in nature and the extent of chimera ranges from few florets of a flower to portion of branch. Appearance of flower colour/shape mutation due to gamma irradiation or as natural sport is in conformity with the earlier published results on other cultivars of chrysanthemum and other ornamentals (Datta 1988, 1997, Datta and Banerji 1990).

Table 14.1: Original and Mutated Ray Floret Colour and Shape

Cultivar	*Original Colour/Shape*	*Mutated Colour/Shape*
Purnima	White	Yellow
Colchi Bahar	Red	Yellow
Maghi	Mauve	Light Mauve, White, Light Yellow, Dark Yellow
Puja	Red Purple/Flat Spoon	YellowOrange/Flat Spoon
		Yellow Orange/Tubular
Kasturba Gandhi	White	Yellow

Swelling of florets of all the cultivars was evident within one week. Shoot buds were initiated within 3 weeks directly from both adaxial and abaxial sides of florets. This direct shoot bud formation from ray floret explants was confirmed from histological and SEM studies. Within 4-6 weeks, numerous shoot buds appeared on the ray floret explants. Elongation of shoot buds occurred rapidly accompanied by initiation and development of new buds in the same medium composition.

Ray floret explants of the cv. Purnima was cultured on all possible combinations of 0.2, 0.5 and 1.0 mg/l BA or kinetin with 0.2, 0.5 and 1.0 mg/l NAA. Direct shoot organogenesis was evident with all the combinations of NAA and BA. The highest percentage of shoot organogenesis as well as number of shoot/explant was obtained on 0.2 mg/l NAA + 0.5 mg/l BA where 60 per cent explants responded with an average of 15 shoots/explant. Increasing the NAA and/or BA level did not improve the shoot organogenic frequency. Among the different NAA and kinetin combinations tested, shoot organogenesis was observed only in 0.2 mg/l NAA + 1.0 mg/l kinetin where 30 per cent explants responded with an average of 7.8 shoots/explant. Rest of the concentrations of kinetin were ineffective.

Ray floret explants of the cv. Colchi Bahar were cultured on 0.2 mg/l NAA and 0.5 mg/l BA. Direct shoot organogenesis was observed in this growth regulator combination where 47.6 per cent explants responded with 6.3 shoots/explant.

Ray floret explants of the cv. Maghi were cultured on 23.23 uM kinetin + 5.37 uM NAA supplemented medium. No shoot bud formation was observed in case of light yellow and dark yellow floret explants while light mauve and white floret explants developed shoot buds at low frequencies (6 and 7 shoots/explant respectively).

In the germplasm collection of NBRI, Lucknow few florets of a flower (cv. Kasturba Gandhi) became yellow due to spontaneous mutation. These mutated yellow ray florets were cultured on 0.2 mg/l NAA + 1 mg/l BA supplemented medium. Shoot buds appeared directly on explant after 21 days of culture initiation. In this medium only 10.25 per cent explants responded with 5 shoot per explant. Rest of the explants produced pale yellow callus. The callus when subcultured on the same medium proliferated but did not produce any shoot bud. However, when the callus was transferred on the induction medium supplemented with 0.2, 0.5, 1.0, 2.0 and 5.0 mg/l GA_3 shoot bud formation was observed. In all these treatments, shoot bud formation was observed in different frequencies and numbers within 2-3 weeks. The best response was obtained on 0.5 mg/l GA_3 treatment where 57.1 per cent callus responded with 9.7 number of shoots/callus. The shoots which were developed directly on the same medium grow slowly with low multiplication rate. When these shoots were subcultured on the same medium but supplemented with 0.5 mg/l GA_3 vigorous growth of shoots were observed.

In all the cultivars, shoots were subcultured on the same induction medium for further growth and multiplication. Shoots of about 2-3 cm in length were excised and cultured on rooting medium. All the shoots in all the cultivars formed roots within one week. Survival of plantlets was 100 per cent when transferred to the field conditions. In the field, plants grew vigorously and produced flowers true-to-floret explant (colour and shape).

The technique of chimera isolation presented here is of immense practical importance for developing new mutants in a relatively short time. This technique could be further extended to other ornamental crops for isolating new and novel flower colour/shape mutants.

Acknowledgements

Thanks are due to Director, NBRI, Lucknow, for providing facilities.

REFERENCES

Chakrabarty, D., Mandal, A.K.A. and Datta, S.K. 1999. Management of chimera through direct shoot regeneration from florets of chrysanrhemum (*Chrysanthemum morifolium* Ramat). J Hort Sci Biotech. 74: 293-296.

Chakrabarty, D., Mandal, A.K.A. and Datta, S.K. 2000. Retrival of new coloured chrysanthemum through organogenesis from sectorial chimeras. Curr Sci., 78: 1060-1061.

D'Amato, F. 1965. Chimera formation in mutagen treated seeds and diplontic selection. In: The Use of Mutations in Plant Breeding, Report of the FAO/IAEA Technical Meeting, Rome 1964. Oxford, Pergmon Press, pp. 302-316.

Index